Catholic Teaching on the Morality of Euthanasia

BY

THE REVEREND JOSEPH V. SULLIVAN, S.T.L.

NIHIL OBSTAT:

Francis J. Connell, C.SS.R., S.T.D.

Censor Deputatus

March 25, 1949

IMPRIMATUR:

✠ Edwin V. O'Hara, D.D.

Bishop of Kansas City, Mo.

March 25, 1949

TO
MY MOTHER AND FATHER

TABLE OF CONTENTS

TABLE OF CONTENTS (Continued)

TABLE OF CONTENTS (Continued)

INTRODUCTION

This brochure represents a part of a doctoral dissertation, the whole of which may be found in the Library of the Catholic University of America in Washington, D. C. Within the limits of this brief work we have attempted to treat the principal points of Catholic moral teaching on the rather modern subject of euthanasia. Very few theologians have treated this subject expressly to any great extent. Those who have done so limit themselves to a few general statements condemning the practice of mercy killing. The main objective of this dissertation is to demonstrate that euthanasia as ordinarily understood today (mercy killing), whether voluntary or compulsory, is immoral. This objective is arrived at by showing that euthanasia is incompatible with God's absolute dominion over life, contrary to the traditional conscience of Western civilization, fraught with disaster in the effects that will inevitably follow from any concession in this matter, and opposed to man's innate desire to live.

The secondary objective is to investigate therapeutic euthanasia (the use of narcotics for the dying) and to determine its morality in the light of Catholic moral teaching. Lastly a brief study is made of the means of prolonging life that should be used relative to the circumstances of the individual case.

My obligations of gratitude are many. I take this opportunity, then, of thanking, first of all, His Excellency, the Most Reverend Edwin V. O'Hara, D.D., Bishop of Kansas City, who gave me the privilege of pursuing graduate studies. I am also deeply grateful to the professors whose classes I attended at the Catholic University of America. In particular I express sincere thanks to the Reverend Francis J. Connell, C.Ss.R., S.T.D., under whose direction the present dissertation was written and to the Reverend Alfred Rush, C.Ss.R., S.T.D., and the Reverend Thomas Martin, S.T.D., readers. To Monsignor John M. Cooper of the School of Anthropology of the Catholic University of America, I am deeply indebted for help in the preparation of the second chapter

Introduction

of this dissertation. Finally I wish to thank four good friends who assisted me morally and materially in the preparation of this work: Very Reverend George W. King of the Diocese of Kansas City; Reverend Julius A. Dorszynski of the Archdiocese of Milwaukee; Mr. John S. Dugan of the Archdiocese of Indianapolis; and Mr. Joseph M. Mills of the Diocese of Owensboro.

JOSEPH V. SULLIVAN, S.T.L.

CHAPTER I

NOTIONS AND DIVISIONS OF EUTHANASIA

INTRODUCTION

The word, *euthanasia,* derived from the Greek (εὐθανασία, f.ĕu-well and θάνατος death), means a gentle and an easy death.[1] As far as is known, the word appeared for the first time in an anglicized form, *euthanasy,* in the year 1636.[2] The meanings given the word have been various both in the Greek and in the English. This dissertation, however, is concerned, not with the etymological meanings of the word *euthanasia,* but rather with the practical aspects of euthanasia itself. From this standpoint a study of the subject will reveal four possible meanings of the word: *natural euthanasia, moral euthanasia, therapeutic euthanasia,* and *merciful euthanasia.* The main concern in this dissertation is the last, *merciful euthanasia,* which in popular language is known as *mercy killing.*[3] In this chapter each division is presented so as to define it and to indicate clearly in what way it differs from the other three.

NATURAL EUTHANASIA

This sense follows from the literal meaning of the term itself, *good death.* Natural euthanasia is simply a natural good death, a painless death that comes upon one without the necessity of medicine or an attending physician to alleviate physical suffering. A recent work of medical ethics treats briefly of this euthanasia and defines it. "It [euthanasia] might be used to denote a quiet, pain-

---

[1] H. Liddel-R. Scott, *Greek-English Lexicon* (New York: Harper & Brothers, 1858), I, p. 567.

[2] *A New English Dictionary on Historical Principles* (Oxford: Clarendon Press, 1897), III, p. 325.

[3] C. Millard, *The Case for Voluntary Euthanasia* (Leicester: Gilroes Co., 1947), p. 1.

1

less death, willed by God to be so, with no interference by medical science."[4]

MORAL EUTHANASIA

This euthanasia can best be characterized by the word, courage. It is not an easy death nor in any sense a gentle death, but rather a good death in the sense of a noble death. Though many are granted natural euthanasia, yet many others are called upon to undergo a painful death, and of them moral euthanasia is expected. Moral euthanasia has been defined as "A valiant facing and accepting of death, a courageous death."[5]

THERAPEUTIC EUTHANASIA

Of all forms of euthanasia, therapeutic euthanasia has received the most attention. It is the only kind of euthanasia treated to any great extent by theologians. Father Bonnar, O.F.M., in speaking on the matter states: "Theologians use the term 'Euthanasia' in the correct sense."[6] He then indicates that he is speaking of what we have called therapeutic euthanasia.[7] Father Hilary Werts, S.J., defines this type of euthanasia as: "the easing of the pain which often accompanies death by the use of therapeutic doses of narcotics."[8] A therapeutic dose is a non-lethal dose. The difference then between therapeutic euthanasia and natural euthanasia is that the former is a death made painless with the help of medical science, while the latter is a painless death without the help of medical science.

MERCIFUL EUTHANASIA

In the present day sense of the term, *euthanasia* signifies, not an easy death, but rather a "merciful release" accomplished in a

---

[4] S. La Rochelle-C. Fink, *Handbook of Medical Ethics* (Westminster: Newman, 1943), p. 165.

[5] *Ibid.*

[6] A. Bonnar, O.F.M., *The Catholic Doctor* (New York: Kenedy & Sons, 1941), p. 98.

[7] *Ibid.*

[8] H. Werts, S.J., "Moral Aspects of Euthanasia," *The Linacre Quarterly,* XIV (1947), p. 27.

painless manner. The only immediate effect of this euthanasia is death. It is defined as "the termination of human life by painless means for the purpose of ending severe physical suffering."[9] This definition was formulated by the Euthanasia Society of America, Inc., with headquarters in New York City and is very similar to the definition given by a sister euthanasia society in England.[10]

Merciful euthanasia may be voluntary or compulsory. It is voluntary when administered only upon the request of the patient. It is compulsory when applied without the consent of the patient, either when he is positively unwilling or when he is incapable of making a decision as in the case of monstrosities, mental defectives and the insane.[11]

---

[9] *Proposed Bill to Legalize Euthanasia,* Article 15, Sect. 300; cf. *infra,* Ch. 3, p. 25 of this dissertation.

[10] *Voluntary Euthanasia Bill,* Sect. 1. This Bill was introduced by Lord Ponsonby in 1936 and the definition given euthanasia reads: "The termination of life by painless means for the purpose of avoiding unnecessary suffering."

[11] A. Wolbarst, "A Doctor Looks at Euthanasia," *Medical Record,* 149 (1939), pp. 354 f.

CHAPTER II

HISTORICAL ASPECTS OF EUTHANASIA

1. INTRODUCTION

Euthanasia (mercy killing) is not a new idea. There are records of the practice of a quasi-euthanasia in the distant past. Anthropological data indicate many approximations to mercy killing, not only among non-civilized communities, but also among the civilizations of Greece and Rome. Not always has the manner of the death been merciful, nor has a doctor or tribal medicine man always been present to administer the death, but in many ways these killings were akin to the present practice of mercy killing.

2. EUTHANASIA IN PRIMITIVE SOCIETIES

Actual neglect, or even abandonment, has been a rather common mode of treatment of the aged in many primitive tribes. A recent anthropological study of 39 native tribes on which definite information could be obtained, indicates that such neglect and abandonment was customary in 18.[1] Abandonment of the aged was well known among the nomadic tribes of the North American Indian as well as among the Hopi, who were known for the high premium they placed on old age.[2] The Omaha were respectful to their aged, and out of fear lest the god (Wakanda) punish them, they would not abandon the sick on the prairie, but the very feeble were left at a camp site, provided with shelter, food, water, and a fire. Thus the aged would live at least a few weeks before starvation.[3] The Kutenai abandoned their aged who could not follow when they

---

[1] L. Simmons, *The Role of the Aged in Primitive Society* (London: Oxford, 1945), p. 225.

[2] *Ibid.*

[3] J. Dorsey, "Omaha Sociology," *Annual Reports of the Bureau of American Ethnology*, III (1882), p. 247.

moved from place to place in search of game and fish, and this
abandonment was not considered disrespectful even by the aged
themselves.[4] And a Crow Indian stated that abandonment was an
approved custom of his people:

> I have heard my grandmother tell of the days when old
> women, too worn out and weak to travel afoot . . . had
> to be left to die. She told me that when an old woman
> was used up, no good any more, the people set up a lodge
> for her, gave her meat and wood for a fire, and then left
> her there to finally die. They could do nothing else.[5]

In Africa the Bushmen forsook their aged and sick when moving camp. There was no hope that they would ever see them again
alive.

> A weak, old, or sick woman is often left behind without
> more ado. A bowl of water, a root or two, a bit of meat,
> are placed beside her ; and the wild beasts soon accomplish
> her destiny.[6]

The Xosa tribe of Africa would carry their helpless aged to the
bush to die, but the extremely frail old women were ignored and
left to die anywhere.

> The natives are respectful to men in old age; but old
> women are excluded from such attention, so that their
> lot is sad indeed. . . . They are frequently left to starve
> or die of exposure and neglect, for they have to depend
> on the charity of the young and strong, and that is a
> precious source of supply to count upon in heathen
> kraals. The one weapon they have is their curse, which
> is a thing greatly dreaded by the people. But to be set
> off against this is that old women are frequently accused
> of being witches . . . these poor old creatures are slowly

---

[4] H. Bancroft, *The Native Races* (New York: Harper & Brothers, 1883),
p. 279.

[5] F. Linderman, *Red Mother* (New York: Longmans, Green & Co., 1932),
p. 82.

[6] F. Ratzel, *The History of Mankind* (London: Macmillan, 1897), Vol.
II, p. 275.

burnt to death because they will not give up the medicine which they are supposed to have used to cause sickness among the cattle or people in the kraal.[7]

A similar practice existed among the Witoto in South America where a quasi-euthanasia was sometimes compulsory. If a man could no longer hunt or fight he was regarded as useless and a mere burden on the community. If he showed no signs of eventual recovery, his friends unhesitatingly gave him up to die, or if a medicine man had not been commissioned to put him out of the way, he was driven into the bush, where the same end was speedily attained. This was done not only to invalids, but also to aged members of the tribe.[8] Another tribe, the Arawak, also of South America, confined their aged in some small corner of the house where they were left alone without food or care.[9]

This anthropological information makes it evident that neglect and abandonment of the aged, the useless and the incurably ill, played a large part in the elimination of the unfit in primitive cultures. Together with abandonment of the aged there existed exposure of ill-born infants and infanticide. Among the non-civilized natives of the Hawaiian Islands infanticide was common.

Meanwhile, over all the island world, abortion and infanticide prevailed. On coral atolls, where the danger was most plainly obvious, these were enforced by law and sanctioned by punishment. On Vaitupu, in the Ellices, only two children were allowed to a couple; on Nukufetau, but one. On the latter the punishment was by fine; and it is related that the fine was sometimes paid, and the child spared.[10]

There is another account of the South Seas and the native customs where many acts of infant exposure are recorded. Though

---

[7] D. Kidd, *The Essential Kafir* (London: Black, 1925), p. 22.

[8] T. Whiffen, *The North-West Amazons* (London: Constable & Co., 1915), p. 170.

[9] W. Roth, "An Inquiry into the Animism and Folk-Lore of the Guiana Indians," *ARBAE*, XXXVIII (1917), p. 702.

[10] B. Stevenson, *In the South Seas* (New York: Scribner's Sons, 1918), p. 38.

superstition is sometimes the cause of infant exposure yet more often some practical reason is present, such as scarcity of food.[11] The natives of Australia used to bury with a dead mother her living infant child. When the body of the mother was placed in the grave, the father himself placed the living child in it with the mother, and having laid the child down, he threw upon it a large stone, and the grave was instantly filled in by the other natives.[12]

3. EUTHANASIA IN GREECE

There was a philosophy of euthanasia in ancient Greece. This is evident from the classic writings of Plutarch, Plato, and Aristotle. These men advocated a quasi-euthanasia. It is interesting to note how men of such high natural culture justify euthanasia from the standpoint of natural ethics. These men condemn suicide and homicide, and yet they view euthanasia as not only permissible but under certain circumstances as the ideal. From their writings it is evident that infanticide was very common in some of the communities of ancient Greece.

> The offspring was not reared at the will of the father, but was taken and carried by him to a place called Lesche, where the elders of the tribes officially examined the infants, and if it were well-built and sturdy, they ordered the father to rear it, and assigned it one of the nine thousand lots of land; but if it was ill-born and deformed, they sent it to the so-called Apothetae, a chasm-like place at the foot of Mount Taygetas in the conviction that the life of that which nature had not well equipped at the very beginning for health and strength, was of no advantage either to itself or the state.[13]

---

[11] W. Barrett, "Notes on the Customs and Beliefs of the Wa-Giriama," *Journal of the Royal Anthropological Institute,* XLI (1911), p. 24.

[12] L. Krzywicki, *Primitive Society and its Vital Statistics* (London: Macmillan, 1934), p. 127.

[13] Plutarch, *Plutarch's Lives* XVI (LCL I, p. 255 Perin): London, Macmillan, 1914.

Plato affirms this practice of infanticide and appears to accept it as the normal thing among the civilized Greeks.

> The offspring of the good, I suppose, they will take to the pen or creche, to certain nurses who live apart in a quarter of the city, but the offspring of the inferior, and of these of the other sort who are born defective, they will properly dispose of in secret, so that no one will know what has become of them.[14]

It is interesting to note that Aristotle even advocated compulsory euthanasia for deformed children. The Philosopher considered that an ideal state should have only fit subjects.

> As to exposing or rearing the children born, let there be a law that no deformed child shall be reared; but on the ground of the number of the children, if the regular customs hinder any of those being exposed, there must be a limit fixed to the procreation of offspring.[15]

Greece seems to have had euthanasia for the aged also. There is indication that at one time there was a law ordering those who were over sixty years of age to drink hemlock, in order that the food might be sufficient for the rest.[16] Once when the Athenians were being besieged by an enemy, they voted, setting a definite age, that the oldest among them be put to death. This siege was raised and there is no indication that the oldest were made to undergo death.[17] It does not appear, however, that a euthanasia for the aged Greeks was nearly so common as it was among the primitive tribes of Africa, South America, or even among the tribes of the North American Indians.[18]

---

[14] Plato, *The Republic* V, IX (LCL I, p. 463 Shorey): New York, Putnam's Sons, 1930.

[15] Aristotle, *Politics* VII, 15, 10 (LCL I, p. 622 Rackham): New York, Putnam's Sons, 1932.

[16] Strabo, *Geography* X, 5, 6 (LCL V, p. 169 Jones): New York, Putnam's Sons, 1928.

[17] *Ibid.*

[18] L. Simmons, *op. cit.*, p. 226.

4. EUTHANASIA IN ROME

With regard to Rome, there is very little to add to the data on euthanasia. The philosophy of the Romans came from the Greeks. Hence the thoughts of Plato and Aristotle would naturally guide the Romans as it did many of the Greeks. Epicurean philosophy was likewise very common in Rome. This philosophy of pleasure forms a natural basis upon which a practical euthanasia could arise. Then the Romans possessed the writings of Seneca, who with great clarity advocated the easy death, euthanasia.

There is no single statement in the writings of Epicurus that would indicate that he advocated euthanasia. He is important simply because his teachings concerning man's end in this life formed a logical basis for euthanasia in that they propose pleasure as the beginning and the end of every action. It would seem then to be in line with this teaching to end life if pleasure could no longer be had in living. If pleasure is all, and there is no pleasure, only pain remaining, the Roman might ask the further reason for his worldly existence. It is evident then that the philosophy of Epicurus would lead the way for the acceptance of the practice of euthanasia.

> For the end of all our actions is to be free from pain and fear, and once we have attained this, the tempest of the soul is laid; seeing that the living creatures have no need to go in search of something that is lacking, nor to look for anything else by which the good of the soul and of the body will be fulfilled. When we are pained because of the absence of pleasure, then, and then only do we feel the need of pleasure. Wherefore we call pleasure the alpha and the omega of a blessed life. Pleasure is our first and kindred good. It is the starting point of every choice and of every aversion, and to it we come back, inasmuch as we make feeling the rule by which to judge of every good thing, and since pleasure is our first and native good, for that reason we do not choose every pleasure whatsoever, but ofttimes pass over many pleasures when a greater annoyance ensues from them.[19]

---

[19] Laertius, Lives of *Eminent Philosophers* X, 128 (LCL II, p. 653 Hicks): New York, Putnam's Sons, 1925.

Then, the thoughts of Epicurus on death would only confirm the Roman in his desire to take the easy way out.

> Accustom thyself to believe that death is nothing to us, for good and evil imply sentience, and death is the privation of all sentience; therefore a right understanding that death is nothing to us makes the mortality of life enjoyable, not by adding to life an illimitable time, but by taking away the yearning after immortality. For life has no terrors for him who has thoroughly apprehended that there are no terrors for him in ceasing to live. Foolish, therefore, is the man who says that he fears death, not because it will pain when it comes, but because it pains in the prospect. Whatever causes no annoyance when it is present, causes only groundless pain in the expectation. Death, therefore, the most awful of evils, is nothing to us, seeing that, when we are, death is not come, and when death is come, we are not. It is nothing then, either to the living or the dead, for with the living it is not, and the dead exist no longer.[20]

Of all the Roman classic writers Seneca is the prototype of the modern advocate of euthanasia. He states clearly that the easy death, the euthanasia, should be chosen when one knows that he is about to die. There is great similarity between his statement on this matter and the statements made by many present day proponents of mercy killing.

> If one death is accompanied by torture, and the other is simple and easy, why not snatch the latter? Just as I shall select my ship when I am about to go on a voyage, or my house when I propose to take a residence, so shall I choose my death when I am about to depart from life. Moreover just as a long-drawn-out life does not necessarily mean a better one, so a long-drawn-out death necessarily means a worse one. There is no occasion when the soul should be humoured more than at the moment of death. Let the soul depart as it feels itself impelled to go; whether it seeks the sword, or the halter, or some draught that attacks the veins, let it proceed and burst the bounds of slavery. Everyone ought to make his

---

[20] *Ibid.*, p. 650.

life acceptable to others besides himself, but his death to himself alone. The best death is the one we like.[21]

In the next chapter of this dissertation reference will be made to many statements of the modern advocates of euthanasia.

5. EUTHANASIA IN NORTHERN EUROPE AT THE CLOSE OF THE NINETEENTH CENTURY

Anthropological records of this period indicate that it was the custom among the Lapps, when moving camp, to abandon the very old and the very ill. They could not carry the disabled persons of the community on the long journey that they had to make. Consequently, those unable to make the journey had to be left behind. The aged were left in a little hut on the mountain and provided with good food and with whatever else could be given them. When this food gave out, however, the aged had the alternative of following to the new camp site or else dying entirely alone. The Lapp looked upon this state of affairs as a sad necessity but not as a sign of cruelty or ingratitude.[22]

6. CONCLUSION

The available data, although sketchy, indicate that a quasi-euthanasia has existed from primitive times to our own. There is no doubt that neglect and abandonment have played a large part in the elimination of aged and enfeebled persons. The same may be said of infanticide and exposure, with regard to the very young. Statistics indicate also that this quasi-euthanasia is more common among cultural conditions of shifting residence. Hence

---

[21] Seneca, *Epistulae Morales* LXX, 11 (LCL II, p. 63 Gummere) : New York, Putnam's Sons, 1920. "Si altera mors cum tormento, altera simplex et facilis est, quidni huic inicienda sit manus? Quemadmodum navem eligam navigaturus et domum habitaturus, sic mortem exiturus e vita. Praeterea, quemadmodum non utique melior est longior vita, sic peior est utique mors longior. In nulla re magis quam in morte morem animo gerere debemus. Exeat, qua impetum cepit; sive ferrum appetit sive laqueum sive aliquam potionem venas occupantem, pergat et vincula servitutis abrumpat. Vitam et aliis adprobare quisque debet, mortem sibi. Optima est, quae placet."

[22] J. Friis, *A Tale of Finmark* (London : Oxford, 1888), p. 64.

it is found often among tribes of hunters, collectors, and herders. In cultural conditions where residence has been permanent, i.e., among farmers and fishers, there is little practice of abandonment of the aged.[28] It is interesting to note that among these primitive tribes the quasi-euthanasia was the result of what they considered an inescapable necessity for the common good. They did not inflict this treatment for the sake of the victim and it was in almost every case compulsory. The reason was lack of food or of means of transportation. It seems that there were entirely different motives behind the quasi-euthanasia of Greece and Rome. Infanticide was seldom resorted to save in the case of deformed offspring. For the most part indications are that the euthanasia of Greece and Rome in the case of adults was voluntary. When applied to the aged the methods used were less crude than those used in primitive cultures.

---

[28] L. Simmons, *op. cit.*, p. 228.

CHAPTER III

The Modern Movement for Euthanasia

1. euthanasia today

At the present time there is a widespread movement in the United States and in Europe to legalize euthanasia, especially voluntary euthanasia. In America a society was formed for this purpose.

> This Society was organized in New York City in January, 1938. Its objects are as follows: (a) By means of an educational campaign to create public demand for the legalization of voluntary euthanasia. (b) To secure the enactment of State laws permitting voluntary euthanasia with a procedure as simple as is consistent with security against abuse.[1]

Prior to 1938, at least in the United States, there was no organization to advance euthanasia. However, a keen interest in the subject was manifested as early as 1925, at the time of the Jean Zinowski mercy killing.

> News items of the past few weeks have featured instances where pity for suffering, which it appeared could be ended only by death, has led several people here and abroad to kill their loved ones, who were in agony.[2]

Jean Zinowski was killed in a Paris hospital by a young Polish actress, Salauislaw Uminska. She admitted the killing and said she did it through mercy and pity for his sufferings.

> The jury acquitted her in three minutes. Now we are told, Paris is realizing the truth of the legal argument that

---

[1] C. Millard, *The Case for Voluntary Euthanasia* (Leicester: Gilroes Co., 1947), p. 6.

[2] *Literary Digest* (Mar. 21, 1925), p. 33.

there is a danger in the precedent of acquittal, for closely upon the Uminska crime of Charity came a second. Anna Virginia Levasseur, a dressmaker, shot and killed her sister, who was suffering from a disease doctors had pronounced 'incurable.' Not long afterwards a similar 'pity murder' occurred in Cresco, Iowa. Will Dunn, a West Point graduate shot and killed his aged parents because they were a 'burden' to themselves and to their relatives, and then ended his own life.[3]

The *New York Times Index* carries news items on these mercy killings;[4] and it is interesting to note that the *New York Times Index* has items concerning euthanasia in nearly every volume since 1920, indicating how important a subject euthanasia was even prior to the present euthanasia movement. Within the last ten years the case study of each mercy killing has been very complete in the *Index*.[5] A further indication of the growing importance of euthanasia is manifested in the following three facts:

Mercy killings now occur in the United States at the rate of one a week;
Mercy killers are almost never convicted;
The stiffest penalty imposed in recent years was a three month prison sentence.[6]

In 1939, the Greenfield case was brought to the attention of the American public. Mr. Greenfield admitted chloroforming his imbecile son, Jerome, but nevertheless pleaded "not guilty" to the charge of murder. The trial ended and Mr. Greenfield was released "without punishment."[7] This case caused great interest throughout the nation in euthanasia, and provoked a Gallup poll. The poll revealed that 46% of the people living in the United States were in favor of some kind of legal euthanasia. A similar poll was taken in Great Britain and the results indicated that 68% of the population would wish euthanasia legalized.[8]

---

[3] *Ibid.*
[4] *New York Times Index* (1925), p. 1210.
[5] *New York Times Index* (1938-1948).
[6] *Time* (Jan. 23, 1939), p. 24.
[7] *New York Times Index* (1939), p. 1414.
[8] *Ibid.*

A growing interest in euthanasia is likewise obvious from the many articles written today in its favor. The trend seems to be toward accepting euthanasia as a part of Western culture. The thoughts of Sherwood Anderson would indicate this.

> 'Not life but the good life.' Didn't Socrates say something like that? It seems to me moral nonsense to go on, in old age, in pain, in uselessness to others . . . to say nothing of self.
> I have always wondered about this whole notion. 'Why . . . if you take your own life you won't go to Heaven.'
> But it's this Heaven that I must wonder about.
> What monstrous egotism . . . that I, who have lived as I have, so often cruel and brutal to others, selfish, self-centered, only occasionally losing self, becoming impersonal, only at rare moments doing any work that means anything to others—
> The thought that this life of mine should be perpetuated, go on forever.
> By what terrible mischance does it deserve that?
> I think, I must think, that it is past all words moral nonsense to say that I must go on, under such circumstances, bringing that much more evil into others' lives.
> For to me disease is evil, old age, decrepitude, with incurable disease the final evil.
> Spring coming, walks in the forest, love, comradeship . . . these all gone.
> Who took these from me?
> Thanks be to those, scientists or others, who have invented or discovered these poisons—perhaps for an almost quiet exit, the door somewhat softly opened.
> An end to my being a nuisance to others.[9]

These remarks by Mr. Anderson were, as he states in his article, motivated by the suicide of his friend, Charlotte Perkins Gilman, who for several years had been in bad health. Miss Gilman was a great advocate of euthanasia, and left a dying message explaining why euthanasia was proper in her case.

> At last duty. Human life consists in mutual service. No grief, pain, misfortune or 'broken heart' is excuse for cut-

---

[9] S. Anderson, "Dinner at Thessaly," *The Forum*, XCV (1936), p. 40.

ting off one's life while any power of service remains. But when all usefulness is over, when one is assured of an imminent and unavoidable death, it is the simplest of human rights to choose a quick and easy death in place of a slow and horrible one.

Public opinion is changing on this subject. The time is approaching when we shall consider it abhorrent to our civilization to allow a human being to lie in prolonged agony which we should mercifully end in any other creature. Believing this choice to be of social service in promoting wiser views on this question, I have preferred chloroform to cancer.[10]

Interest in euthanasia has increased as man's life expectancy has increased. In 1900, when the average person could expect to live to be about 49, such painful deaths as cancer were not as common as today.[11] Now the average person can expect to live to the age of 66, and a population growing older means more cancer cases. Reports indicate that more than one-fifth of the babies born today will become victims of cancer sometime before they die.[12] An increase in cancer deaths means an increase in painful deaths. That is one reason why euthanasia is gaining in public favor.

Of the 172,700 persons who died of cancer in the United States in 1945 alone, do you suppose all were permitted to drain the last bitter drop of the cup of pain? Many of them had faithful friends who helped them to die in dignity and repose instead of shrieking, groaning, and curseing 'til their breath failed.

But such friends risked their own lives in doing the decent and loving thing. They could have been tried and convicted of murder for obeying the beatitude, 'Blessed are the merciful, for they shall obtain mercy.'[13]

---

[10] R. Howard, "Taking Life Legally," *Magazine Digest*, XC (1947), p. 33.

[11] D. Johnson, *Facing the Facts about Cancer* (New York: Committee Press, 1947), p. 6.

[12] *Ibid.*

[13] C. Potter, "The Case for Euthanasia," *Readers' Scope*, XXX (1947), p. 113.

What is true of cancer is true of many other diseases that may come in later life. Painful diseases are on the increase after middle age. And with a greater number yearly reaching the age of 70 it stands to reason that yearly painful deaths will be on the increase. Unless research reveals a more successful cure of these diseases, the death rate due to them will continue to rise.[14] Euthanasia, therefore, does have a practical appeal in these days of sanitary controls, vaccines, sulfa drugs, penicillin, and better trained physicians, for as a result of all this modern science, more people are living to an old age.

Interest in euthanasia is also provoked out of motives for the common good. There are indications that some favor the elimination of the unfit for the sake of society.

> The question of legal euthanasia is often brought up in connection with the mentally defective and the insane. The State of New York alone spends about thirty million dollars a year for the care of those incurables. One may well ask what useful or human purpose is served by keeping these unfortunates alive for many years. Undoubtedly the only point in keeping them alive is their legal right to live, with which is associated the faint hope that some of them may get well. Records show that very few get well and fewer stay well.[15]

Euthanasia in such cases would be compulsory. There are but few, as yet, who advocate compulsory euthanasia. The chief objection is that the physician might make a wrong judgment and send some curable person to death. Even the advocates themselves are willing to admit that possibility.

> Doctors are only human beings, with few if any supermen among them. They make honest mistakes, like other men, because of the limitations of the human mind. They might conceivably agree on legal euthanasia in a certain case, only to find on autopsy, that they had made a wrong diagnosis. I mention these matters because of the difficulty involved in the decision as to incurability, etc.[16]

---

[14] D. Johnson, *op. cit.*
[15] A. Wolbarst, "The Right to Die," *The Forum*, XCIV (1935), p. 332.
[16] *Ibid.*

But over and above any specific objections to the practice, there is another reason why compulsory euthanasia has received so little public support. And it is simply that compulsory euthanasia does not have a strong, active society behind it, such as voluntary euthanasia has. In other words, compulsory euthanasia stands today where voluntary euthanasia stood in 1925.

2. THE PRESENT ATTEMPT AT LEGISLATION

Prior to the last decade there was no attempt to legalize euthanasia in any State of the Union. News items of the twenties and early thirties indicate opposition to such a move, and even the liberal opinion of that time considered mercy killing dangerous.

> That there have been cases where physicians recognized the hopelessness of life and permitted, if they did not hasten, the quick coming of death is certain—but no general rule for such conduct might well be formulated. And that action comparable with that of the woman just freed by the Paris court has been taken before by individuals for the sake of loved ones is unquestioned. Yet the state would not do well if such actions were permitted when known to go by default without at least some formal hearing in the matter. Human life when all its possible joys have fled, is an extremely valueless thing to the possessor. Yet it is so priceless a thing if there be but a spark of hope in it that its taking must be considered something of importance.[17]

It was considered a matter of grave doubt whether the point would ever be reached where society would permit euthanasia. The common fear was that if the bars were broken down, this power over life would often be used recklessly and society would suffer.[18]

Today, however, many of the public think differently. The movement for voluntary euthanasia has gained many followers. The Euthanasia Society has recently sent inquiries to all the physicians of New York State seeking their views on the subject. Of

[17] *Literary Digest* (March 21, 1925), p. 33.
[18] *Ibid.*

the three thousand two hundred and seventy-two replies, 80% were favorable.[19] In addition to this fact, a special committee of physicians was formed to further the movement for legal euthanasia.

> No wonder, indeed, that within the last few months 1,769 physicians in New York State have joined our Committee of Physicians for the Legislation of Voluntary Euthanasia.[20]

Some of the physicians stated their mind on the subject publicly. Dr. Hinman believes that euthanasia, by which he means allowing an incurable sufferer to die, is practiced daily by physicians. For this reason he thinks euthanasia should be legalized.[21] Dr. Foster Kennedy is in favor of euthanasia especially for the "helpless ones who should never have been born." He is opposed to voluntary euthanasia as now proposed by the Euthanasia Society.[22]

The movement for voluntary euthanasia has likewise gained support from many clergymen of New York. A statement was prepared by the Euthanasia Society and submitted to many ministers for signature. So far fifty clergymen have signed this proposed statement.[23]

> A proposal has been put forward to legalize voluntary euthanasia, i.e., painless death for persons desiring it, who are suffering from incurable, fatal and painful disease. A Bill has been drafted to give effect to this, and the proposal is receiving encouragement and support from many thinking people.
> Such a proposal raises important issues on ethical, legal and medical grounds. As regard the ethical issue, we, the undersigned, after giving the matter careful consideration, wish to state that, in our opinion, voluntary euthanasia, under the circumstances mentioned above, should

[19] R. Roberts, *Merciful Release* (New York: Euthanasia Press, 1947), p. 6.
[20] C. Potter, "The Case for Euthanasia," *Readers Scope*, XXX (1947), p. 113.
[21] F. Hinman, "Power Over Life and Death," *Journal of Nervous and Mental Diseases*, IC (1944), p. 645.
[22] F. Kennedy, "Unfit to Live," *American Journal of Psychiatry*, IC (1942), p. 13.
[23] Euthanasia Society of America, *Special Report* (1947)

not be regarded as contrary to the principles of morality or religion.[24]

Dr. Charles Breck Ackley, St. Mary's Manhattanville Church, N. Y. C. (Prot. Episcopal).

Rev. L. M. Birkhead, Nat'l Director, Friends of Democracy, Inc., N. Y. C.

Rev. Shelton Hale Bishop, St. Philip's Church, N. Y. C. (Episcopal).

Dr. A. Lynn Booth, Church of the Reconciliation, Utica, N. Y. (Unity).

Dr. W. Russell Bowie, Union Theol. Sem., N. Y. C.

Rev. Clarence E. Boyer, Adam-Parkhurst Memorial Church, N. Y. C. (Presbyterian).

Rev. Weston A. Cate, First Universalist Church, Rochester, N. Y.

Rev. Karl M. Choworowsky, Flatbush Unitarian Church, Brooklyn, N. Y.

Rev. J. Henry Carpenter, Brooklyn Church and Mission Federation, Brooklyn, N. Y.

Dr. Henry Sloan Coffin, President, Union Theol. Sem., N. Y. C.

Rev. Dale DeWitt, Regional Director, American Unitarian Assn. for Middle Atlantic States, N. Y. C.

Rev. I. J. Domas, Christ Church, Middletown, N. Y. (Universalist)

Rev. Samuel M. Dorrance, St. Ann's Church, Brooklyn, N. Y.

Rev. Carleton M. Fisher, Church of the Messiah, Buffalo, N. Y. (Universalist).

Dr. Harry Emerson Fosdick, The Riverside Church, N. Y. C.

Rev. Aron S. Gilmartin, Church of Our Father, Newburg, N. Y. (Unitarian).

Rev. Charles G. Girelius, Unitarian Church, Barneveld, N. Y.

Dr. Cornelius Greenway, All Souls Universalist Church, Brooklyn, N. Y.

Dr. Ivar Hellstrom, The Riverside Church, N. Y. C.

Rev. Ralph K. Hickok, Whytcote, Aurora-on-Cayuga, N. Y.

Dr. Ralph B. Hindman, First Presbyterian Church, Buffalo, N. Y.

Rev. Harry Hooper, Unitarian Church, Staten Island, N. Y.

Rev. Douglas Horton, Gen. Council of Congregational Christian Churches, N. Y. C.

Dr. Murray Shipley Howland, First Presbyterian Church, Binghamton, N. Y.

Dr. J. Donald Johnson, First Unitarian Church, Flushing, L. I., N. Y.

Rev. Algernon D. Black, Leader Society for Ethical Culture, N. Y. C.

Rabbi Max Meyer, Flushing Free Synagogue, N. Y.

Dr. George Paul T. Sargent, Rector, St. Bartholomew's Church, N. Y. C. (Episcopal).

Rev. Wendell Phillips, Christ's Church, Rye, N. Y.

---

[24] "Statement on the Ethical Aspects of Euthanasia by Fifty Religious Leaders of New York State," Euthanasia Society, Jan., 1947, Mimeographed.

Rev. Paul Jones, Union Church of Bay Ridge, Brooklyn, N. Y.

Rev. Spear Knebel, Trinity Episcopal Church, Albany, N. Y.

Dr. John Howland Lathrop, Church of the Saviour, Brooklyn, N. Y.

Rev. Henry Smith Leiper, Universal Christian Council, N. Y. C.

Dr. Clifton Macon, Morningside Prot. Episcopal Church, N. Y. C.

Rev. Elmore M. McKee, St. George's Church, N. Y. C.

Rev. James W. McKnight, Universalist & Unitarian Church, Mt. Vernon, N. Y.

Dr. John Howard Melish, Church of the Holy Trinity, Brooklyn, N. Y.

Dr. William P. Merrill, Pastor Emeritus, Brick Presbyterian Church, N. Y. C.

Rev. William J. Metz, Universalist Church, Central Square, N. Y.

Rev. George E. O'Dell, Secretary, The American Ethical Union, N. Y. C.

Dr. David E. Roberts, Union Theol. Seminary, Dean, N. Y. C.

Rev. Theodore F. Savage, Executive Secretary, Presbytery of N. Y.

Dr. Guy Emery Shipler, Editor, The Churchman, N. Y. C.

Rev. Vincent B. Silliman, Hollis Unitarian Church, Hollis, L. I., N. Y.

Dr. George A. Simons, Glendale Christ Methodist Church, Brooklyn, N. Y.

Dr. Ralph W. Sockman, Christ Church, N. Y. C. (Methodist).

Rev. Norris L. Tibbetts, The Riverside Church, N. Y. C.

Rev. Joseph H. Titus, Grace Church, Jamaica, L. I., N. Y.

Dr. Henry Van Dusen, Dean, Union Theol. Seminary, N. Y. C.

Rev. Clifford H. Vessey, White Plains Community Church, White Plains, N. Y.

Rev. Kenneth C. Walker, First Unitarian Church, Albany, N. Y.

Dr. P. A. Wallace, A.M.E. Zion Church, Brooklyn, N. Y.

Dr. Gerald Watkins, Lake Ave. Baptist Church, Rochester, N. Y.

Rev. Edwin H. Wilson, First Unitarian Society, Schenectady, N. Y.

More recently an even larger group of non-Catholic clergymen have approved of euthanasia.

The Euthanasia Society plans at a later date to gain the signatures of physicians and clergymen throughout the nation. The Society has formed a national body for this purpose, known as the *American Advisory Council*. This body consists of one hundred members including such figures as: Eugene O'Neil, Dr. Harry E. Fosdick, Fannie Hurst, Rev. Theodore Savage, Nebraska State Senator John H. Comstock and Margaret Sanger.[25] This body has also prepared the euthanasia proposal that will be submitted to the New York State Assembly. The proposal is known as: *The Pro-*

— — —

[25] R. Roberts, *op. cit.*, p. 2.

posed Bill to Legalize Euthanasia. At the end of this chapter the entire bill is presented.

These steps have gained prestige for the euthanasia movement and have made legal euthanasia a definite possibility. It is interesting to note that Senator Comstock as early as 1937 introduced into the Nebraska Assembly his own bill for legalized voluntary euthanasia. The bill, however, was never submitted to a vote.[26] If the present euthanasia proposal is submitted to a vote in the New York State Assembly, it will be the first time in American history that euthanasia has been voted upon.

In view of the coming test in the New York Assembly the Euthanasia Society has prepared a public statement for the press.

> We feel deep compassion for those condemned to die. But there are others—and their numbers are growing rapidly with our aging population—who are condemned to live. Helpless, hopeless, tortured, doomed to a death that is slow in coming, they pray for release denied to them by law. Cancer alone, the disease responsible for the greatest number of lingering painful lives, has over a half million known sufferers in the United States today. And it is only one of the degenerative diseases to which our aging population is subject.
>
> Some who face long months—or years—of agony, attempt self-destruction by crude methods. They jump from windows. They slash their wrists. They swallow corrosive disinfectants.
>
> In other cases, a devoted relative puts the sufferer out of his misery, risking public condemnation and conviction for murder.
>
> But under our present archaic laws the great majority of those afflicted with incurable painful diseases who beg for merciful release drag out their intolerable existence until cruel unreasoning nature brings on the end. Society turns a deaf ear to their agonized cries.
>
> In our age of humanity and science is this justifiable?
>
> Is it not our duty to do something about this futile, unnecessary, human suffering? (We do not allow animals in a similar plight to suffer; we help them out of their misery.)

---

[20] *New York Times Index* (1937), p. 1483.

We believe that a solution of the problem can and will be
found when we can clear away the ignorance, misunder-
standing and fear on the subject.
The Euthanasia Society of America, Inc., was formed to
carry on an educational campaign so that the public may
know the truth.
Out of the knowledge, legal measures will be developed
with careful safeguards against abuse—that will make release (euthanasia) available to incurable sufferers.[27]

3. A SUMMARY OF THE EUTHANASIA PROPOSAL

Any sane person over twenty-one years of age, suffering from
severe physical pain caused by a disease for which no remedy
affording lasting relief or recovery is at the time known to medical
science, may address to the court of record a signed and attested
petition for euthanasia, accompanied by an affidavit from his attending physician to the effect that the disease is, in his opinion,
incurable. The Court shall then appoint a commission of three persons, at least two of them doctors, to investigate all the factors involved in the case and to report to the Court whether the patient
understands the purpose of the petition and comes within the
provisions of the act. If a favorable report is made by the commission, the Court shall grant the petition, and euthanasia, if still
requested by the patient, may be administered by a physician or
any other person chosen by the patient or the commission. The
Bill in its present form is permissive and not mandatory.

4. THE RESOLUTION

WHEREAS, large numbers of our population, notwith-
standing the advance of medical science, suffer from painful diseases for which neither prevention, cure, nor lasting relief has been found, and
WHEREAS, the proportion of the aged in our population, who are subject to the painful, chronic, degenerative

---

[27] R. Roberts, *Condemned to Live* (New York: Euthanasia Press, 1947),
p. 2.

disease, is rapidly increasing and death rates from cancer reached a new high in 1946, and

WHEREAS, many incurable sufferers, facing months of agony, attempt crude violent methods of suicide; while in other cases those in attendance upon hopeless incurables who plead for merciful release, secretly put them out of their misery by administering euthanasia and thereby render themselves liable to prosecution as murderers, and

WHEREAS, to permit the termination of useless, hopeless suffering at the request of the sufferer, is in accord with the humane spirit of this age, and many leaders of public opinion recommend that voluntary euthanasia (merciful release petitioned for by the sufferer) should be permitted by law, brought out into the open and safeguarded against abuse, rather than as at present practiced illegally, surreptitiously and without regulation, and

WHEREAS, a Committee of 1,776 Physicians for Legalization of Voluntary Euthanasia in New York State urges the amendment of the law to permit and safeguard against abuse the merciful release of incurable sufferers who plead for it, therefore be it

RESOLVED, (if the Assembly concur) that a joint legislative committee be and hereby is created to make a thorough study and investigation into the practice of voluntary euthanasia of the incurable sufferer in New York State and the law penalizing this practice as murder; in order to determine whether amendment of such law would promote the public welfare and if so, to recommend appropriate legislation, and be it further

RESOLVED, (if the Assembly concur) that such Committee shall consist of three members of the Senate to be appointed by the temporary president of the Senate and four members of the Assembly, to be appointed by the Speaker of the Assembly and be it further

RESOLVED, (if the Assembly concur) that there be and hereby is appropriated and made available from the legislative contingent fund the sum of five thousand dollars ($5,000) or so much thereof as may be necessary to pay the expenses of the Committee hereby created. The money hereby appropriated and made available to the Committee shall be paid on the audit and warrant of the Comptroller on vouchers approved in manner provided by law.[28]

---

[28] "Resolution," Euthanasia Society, Jan., 1947, Mimeographed.

AN ACT

To Amend the Public Health Law, and the Penal Law, in Relation to Voluntary Euthanasia.

——— — —   ——   —————

The People of the State of New York, represented in the Senate and Assembly, do enact as follows:

Section One

Chapter forty-nine of the laws of Nineteen Hundred Nine, entitled, AN ACT IN RELATION TO THE PUBLIC HEALTH, CONSTITUTING CHAPTER FORTY-FIVE OF THE CONSOLIDATED LAWS, is hereby amended by adding thereto a new article, to be article fifteen, to read as follows:

Article 15

Voluntary Euthanasia

Section 300. Definitions.
301. Who may receive euthanasia.
302. Jurisdiction of courts.
303. Application to court.
304. Investigation and report of committee appointed by court.
305. Administration of euthanasia.
306. Immunity from criminal or civil liability.

Sec. 300. Definitions. As used in this article:

"Euthanasia" means the termination of human life by painless means for the purpose of ending severe physical suffering.

"Patient" means the person desiring to receive euthanasia.

"Physician" means any person licensed to practice medicine in the State of New York.

Sec. 301. Who may receive euthanasia. Any person of sound mind over twenty-one years of ago who is suffering from severe physical pain caused by a disease for which no remedy affording lasting relief or recovery is at the time known to medical science may have euthanasia administered.

The desire to anticipate death by euthanasia under these conditions shall not be deemed to indicate mental impairment.

Sec. 302. JURISDICTION OF COURTS. Any justice of the Supreme Court of the judicial district, in which the patient resides or may be, or any other judge of a county court of any county in which the patient resides or may be, to whom a petition for euthanasia is presented, shall have jurisdiction of and shall grant euthanasia upon the conditions and in conformity with the provisions of this article.

Sec. 303. APPLICATION TO COURT. A petition for euthanasia must be in writing signed by the patient in the presence of two witnesses who must add their signatures and the post-office addresses of their domicile. Such petition must be made in substantially the following form:

To the..Court

I..residing at ....
hereby declare as follows:

I am ....years of age and am suffering severe physical pain caused, as I am advised by my physician, by a disease for which no remedy affording lasting relief or recovery is at this time known to medical science.

I am desirous of anticipating death by euthanasia and hereby petition for permission to receive euthanasia.

The names and addresses of the following persons are as follows or, if unknown to me, I so state:

Father ....
Mother ....
Spouse ....
Children ....
Uncles ....
Aunts ....

Signed....................................

In the presence of

....   residing at ....

....   residing at..................

Date ....

Such petition must be accompanied by a certificate signed by the patient's attending physician in substantially the following form:

To the ....Court

I. .. .. of ....
do hereby certify as follows:

I have attended the patient, ...
since..... ....

It is my opinion and belief that the patient is suffering severe physical pain caused by a disease for which no remedy affording lasting relief or recovery is at the present time known to medical science.

The disease from which the patient is suffering is known as ....

I am satisfied that the patient understands the nature and purpose of the petition in support of which this certificate is issued and that such disease comes within the provisions of section three hundred one of article fifteen of the Public Health Law.

Signature.... .. ..

Date. .Medical Qualifications. ....

If for any reason, the patient is unable to write, he may execute the petition by making his mark which shall be authenticated in the manner provided by law.

Sec. 304. INVESTIGATION AND REPORT OF COMMITTEE APPOINTED BY COURT.

The judge or justice to whom a petition for euthanasia has been presented shall appoint a committee of three competent persons, who are not opposed to euthanasia as herein provided, of whom at least two must be physicians and members of a county or district medical society, who shall forthwith examine the patient and such other persons as they deem advisable or as the court may direct and within five days after their appointment, shall report to the court whether or not the patient understands the nature and purpose of the petition and comes within the provisions of section three hundred one of this article. The court must either grant or deny the petition within three days of its receipt.

If the said committee shall report in the affirmative the court shall grant the petition unless there is reason to believe that the report is erroneous or untrue, in which case the court shall state in writing the reason for denying the petition.

If the petition shall be denied an appeal may be taken to the appellate division of the supreme court, and/or to the Court of Appeals.

Sec. 305. ADMINISTRATION OF EUTHANASIA. When the petition has been granted as herein provided, euthanasia shall be administered in the presence of the committee, or any two members thereof, appointed according to section 304 of this article, by a person chosen by the patient or said committee, or any two members thereof, with the patient's consent; but no person shall be obliged to administer or receive euthanasia against his will.

Sec. 306. IMMUNITY FROM CRIMINAL OR CIVIL LIABILITY.

A person to whom euthanasia has been administered under the conditions of this act shall not be deemed to have died a violent or unnatural death nor shall any physician or person who has administered or assisted in the administration thereof be deemed to have committed any offense criminal or civil, or be liable to any person whatever for the damages or otherwise.

Section Two

The penal law is hereby amended by adding thereto a new section, to be section ten hundred fifty-six, to read as follows:

Sec. 1056. APPLICATION OF ARTICLE TO EUTHANASIA.

Death resulting from euthanasia administered pursuant to and in accordance with the provisions of article fifteen of the public health law shall not constitute a crime or be punishable under any provisions of this act.

Section Three

This act shall take place immediately.

CHAPTER IV

The Morality of Euthanasia

This chapter will discuss the moral aspects of mercy killing, whether voluntary or compulsory. The thesis which the writer will attempt to prove is that it is never lawful for man on his own authority to kill the innocent directly. If the thesis is morally sound, it follows that mercy killing is never permissible.

THESIS: IT IS NEVER LAWFUL FOR MAN ON HIS OWN AUTHORITY TO KILL THE INNOCENT DIRECTLY.

DEFINITION OF TERMS

Man as understood in this thesis includes both an individual member of society, and society itself. Hence the thesis states that not even the state may kill an innocent person directly. *Man* is spoken of in contradistinction to God who may kill the innocent directly either by His own act or through an angel or another man. God has full dominion over life, the life of the good and the life of the wicked. If a man were to kill innocent persons on the authority of God he would do no wrong.

> See ye that I alone am, and there is no other God beside me: I will kill and I will make to live: I will strike, and I will heal, and there is none that can deliver out of my hand.[1]

There are indications in Sacred Scripture of the Lord demanding the life of the innocent. It was the Lord who smote the firstborn of Egypt.

> And it came to pass at midnight, the Lord slew every first-born in the land of Egypt, from the firstborn of the

---

[1] Deuteronomy, 32:39.

29

Pharao, who sat on his throne, unto the firstborn of the captive woman that was in prison. . . .[2]

The Book of Deuteronomy gives an instance where at the command of God even the innocent women and children were put to death.

And the Lord said to me: Fear not: because he is delivered into thy hand, with all his people and his land: and thou shalt do to him as thou hast done to Sehon king of the Amorrhites, that dwell in Hesebon. . . . And we utterly destroyed them, as we had done to Sehon the king of Hesebon, destroying every city, men and women and children.[3]

A similar command was given with regard to the people of Asedoth.

So Josue conquered all the country of the hills and of the south and of the plain, and of Asedoth, with their kings: he left not any remains therein, but slew all that breathed as the Lord the God of Israel had commanded him.[4]

The Old Testament presents other cases where at God's command or request the innocent were put to death.[5] Even though God's command to Abraham that he offer his son Isaac as a holocaust, was only a test,[6] yet God by reason of His full dominion over life, could have required such a sacrifice. It is of interest to note the remark of St. Augustine concerning the obedience of Abraham to this command of God.

Abraham indeed was not merely deemed guiltless of cruelty, but was even applauded for his piety, because he was ready to slay his son in obedience to God, not to his own passion. And it is reasonably enough made a question, whether we are to esteem it to have been in compliance with a command of God that Jephthat killed his daughter because she met him when he had vowed that he

---

[2] Exodus, 12:29.
[3] Deuteronomy, 3:2,6.
[4] Josue, 10:40.
[5] Deuteronomy, 13.
[6] Genesis, 22:11.

would sacrifice to God whatever first met him as he returned victorious from battle.[7]

Today, however, there is no indication that God is giving anyone orders to kill the innocent. As a matter of fact there is not a single instance in the whole New Testament where God commanded one man to kill another innocent man.

The killing spoken of in the thesis refers obviously to the killing of a *human* being. It is *human* life that is sacred and not life *qua* life.[8] If life *qua* life were sacred then it would be wrong for man to destroy any life whether it be vegetative, animal or rational. This distinction is made to escape the menace of a double mistake. The mistake can come, as Father Farrell has said, from a denial of any specific difference between man and the animals. For in the event of such a denial, if no animal life is sacred, there is nothing sacred about the life of man; or proceeding in the other direction, if man's life is sacred, all life is sacred.[9]

It has been common ethical teaching for ages that the life of plants and animals is subject to man, so that man may kill them for his use.

> . . . plants exist for the sake of animals and animals for the good of man, the domestic species both for his service and for his food, and if not all at all events most of the wild ones for the sake of his food and of his supplies of other kinds, in order that they may furnish him both with clothing and with other appliances. If therefore nature makes nothing without purpose or in vain, it follows that nature has made all the animals for the sake of men.[10]

[7] St. Augustine, *De civitate Dei* I, 21 (CSEL 40) ". . . et Abraham non solum non est culpatus crudelitatis crimine, verum etiam laudatus est nomine pietatis, quod voluit filium, nequaquam scelerate, sed obedienter occidere; et merito quaeritur, utrum pro iussu Dei sit habendum, quod Iephte filiam, quae patri occurrit, occidit, cum se vovisset immolaturum Deo, quod ei redeunti de praelio victori primitus occurrisset."

[8] Webster's Collegiate Dictionary (Springfield: Merriam Co., 1940), p. 874. Sacred is to be understood in the sense of inviolable, not to be destroyed.

[9] W. Farrell, *A Companion to the Summa* (New York: Sheed & Ward, 1940), III, p. 195.

[10] Aristotle, *Politics*, I, 3 (LCL I, Rackham): New York: Putnam's Sons, 1932.

Sacred Scripture presents the truth very clearly in the following two statements:

> And he said: Let us make man to our image and likeness: and let him have dominion over the fishes of the sea, and the fowls of the air, and the beasts, and the whole earth, and every creeping creature that moveth upon the earth.[11]

> And God said: Behold I have given you every herb bearing seed upon the earth, and all the trees that have in themselves seed of their own kind, to be your meat.[12]

The traditional Christian view on the sacredness of life is affirmed by St. Augustine, together with the fact that man may kill plants and animals.

> When we hear it said, 'Thou shalt not kill,' we do not take it as referring to trees, for they have no sense, nor to irrational animals, because they have no fellowship with us. Hence it follows that the words, 'Thou shalt not kill' refer to the killing of a man.[13]

St. Thomas in a similar treatment to that of St. Augustine develops this common view of the lawfulness of killing plants and animals.

> Wherefore it is not unlawful if man use plants for the good of animals, and animals for the good of man, as the Philosopher states. Now the most necessary use would seem to consist in the fact that animals use plants, and men use animals for food, and this cannot be done unless these be deprived of life: wherefore it is lawful both to take the life from plants for the use of animals, and from animals for the use of man.[14]

---

[11] Genesis, 1:26.

[12] Genesis, 1:29.

[13] St. Augustine, *De civitate Dei* 120 (CSEL 40). "Cum audimus: Non occides, non accipimus hoc dictum esse de fructetis, quia nullus eis est sensus; nec de irrationalibus animalibus, quia nulla nobis ratione sociantur; restat ergo, ut intelligamus de homine quod dictum est; Non occides."

[14] St. Thomas, *Summa Theologica*, II, II, Q. 64, Art. I. ". . . et ideo si homo utatur plantis ad utilitatem animalium, et animalibus ad utilitatem hominum, non est illicitum, ut patet per Philos. in. i Polit. (cap. 5 et 7):

There is likewise the profession of faith against the doctrine of the Waldenses in which the eating of flesh is declared free from guilt.[15] Hence, theologians unanimously teach that man may kill plants and animals.[16] It is clear therefore that man has dominion over plants and animals and may even kill them for his use. He may not abuse them—for example by unnecessary cruelty to animals—but such abuse is sinful—not because of any rights these creatures have, but because man in so acting, is acting against right reason.[17] Nevertheless, such abuse is not necessarily grave matter.

The term *innocent* in the thesis does not mean that the person is free from all sin. The term *innocent* means only that the person has not been judged worthy of death by a lawful authority, i.e., the state, a world state, a military court, and that he is not a combatant opposing a nation that is fighting a just war, or an unjust aggressor. If he is any of these, then the person is not to be understood as coming under the term *innocent* as used in the thesis.

A criminal therefore may be killed directly by a lawful authority. This is the common ethical teaching and was taught by Aristotle.

> . . . for a bad man will do ten thousand times as much evil as a brute.[18]
> For as man, when perfected, is the best of animals, so he is the worst of all when sundered from the law and justice.
> For unrighteousness is most pernicious when possessed

---

inter alios autem usus maxime necessarius esse videtur, ut animalia plantis utantur in cibum, et homines animalibus; quod sine mortificatione eorum fieri non potest: et ideo licitum est, et plantas mortificare in usum animalium, et animalia in usum hominum."

[15] Innocent III, *Professio fidei Durando de Osca et sociis eius Waldensibus praescripta* (DBU, 425).

[16] J. Aertnys-C. Damen, *Theologia moralis* (Turin: Marietti, 1944), I, nr. 583; H. Noldin-A. Schmitt, *Summa theologiae moralis* (Rome: Pustet Co., 1941), II, nr. 346; H. Davis, *Moral and Pastoral Theology* (New York: Sheed & Ward, 1943), II, p. 258; Gury-Ferreres, *Compendium theologiae moralis* (Barcelona: 1906), II, nr. 56; A. Koch-A. Preuss, *Handbook of Moral Theology* (St. Louis: Herder, 1928), III, p. 59.

[17] M. Cronin, *The Science of Ethics* (New York: Benziger, 1939), II, p. 92.

[18] Aristotle, *Nicomachean Ethics* VII, 6 (LCL I, Jowett), London: Harvard Press, 1940.

of weapons, and man is born possessing weapons for the use of wisdom and virtue which it is possible to employ entirely for the opposite ends. Hence when devoid of virtue man is the most unholy and savage of animals.[19]

The right of a lawful authority to kill an evildoer is likewise a part of the Christian tradition. St. Augustine teaches that though one is guilty of murder if he kills a criminal without authority, yet he may do so when exercising public authority.[20] St. Thomas develops the argument of St. Augustine and approves in certain cases the use of capital punishment.

Now every part is directed to the whole, as imperfect to the perfect, wherefore every part is naturally for the sake of the whole. For this reason we observe that if the health of the whole body demands the excision of a member, through its being decayed or infectious to the other members, it will be both praiseworthy and advantageous to have it cut away. Now every individual person is compared to the whole community, as part to whole. Therefore if a man be dangerous and infectious to the community, on account of some sin, it is praiseworthy and advantageous that he be killed in order to safeguard the common good.[21]
Hence, although it be evil in itself to kill a man so long as he preserve his dignity, yet it may be good to kill a man who has sinned, even as it is to kill a beast.[22]

_____

[19] Aristotle, *Politics* I, 1 (LCL I, p.   Rackham) : New York: Putnam's Sons, 1932.

[20] St. Augustine, *De civitate Dei* I, 21 (CSEL 40).

[21] St. Thomas, *Summa Theologica,* II, II, Q. 64, Art. 2. ":omnis autem pars ordinatur ad totum, ut imperfectum ad perfectum; et ideo omnis pars naturaliter est propter totum; et propter hoc videmus, quod si saluti totius corporis humani expediat praecisio alicujus membri, puta cum est putridum, vel corruptivum aliorum membrorum, laudabiliter, et salubriter abscinditur: quaelibet autem persona singularis comparatur ad totam communitatem, sicut pars ad totum; et ideo si aliquis homo sit periculosus communitati, et corruptivus ipsius propter aliquod peccatum, laudabiliter, et salubriter occiditur, ut bonum commune conservetur:"

[22] St. Thomas, *Summa theologica,* II, II, Q. 64, Art. 2. ". . . et ideo quamvis hominem in sua dignitate manentem occidere sit secundum se malum; tamen hominem peccatorem occidere potest esse bonum, sicut occidere bestiam."

A further indication of this common Christian tradition approving the killing of malefactors, is the profession of faith prepared by Pope Innocent III against the doctrine of the Waldenses. These heretics maintained that it was gravely sinful to kill even criminals condemned to death. The profession of faith makes the Catholic position clear.

> We assert concerning the power of the state, that it is able to exercise a judgment of blood, without mortal sin, provided it proceed to inflict the punishment not in hate, but in judgment, not incautiously, but after consideration.[23]

The traditional teaching is also evidenced by the common teaching of theologians who maintain that a malefactor may be condemned to death by a lawful authority and killed by one appointed by the state.[24] This traditional teaching has the support of Sacred Scripture for both the Old and the New Testament make direct statements on the matter.

> The soul that sinneth, the same shall die.[25]
> Whosoever shall shed man's blood, his blood shall be shed: for man was made to the image of God.[26]
> Neither can it be otherwise expiated, but by his blood that hath shed the blood of another.[27]
> For he is God's minister to thee, for good. But if thou do that which is evil, fear: for he beareth not the sword in vain. For he is God's minister: an avenger to execute wrath upon him that doth evil.[28]

---

[23] Innocent III, *Professio fidei Durando de Osca et sociis eius Waldensibus praescripta* (DBU, 425). "De potestate saeculari asserimus, quod sine peccato mortali potest iudicium sanguinis exercere, dummodo ad inferendam vindictam non odio, sed iudicio, non incaute, sed consulte procedat."

[24] J. Aertnys-C. Damen, *Theologia moralis*, I, nr. 569; H. Noldin-A. Schmitt, *Summa theologiae moralis*, II, nr. 330; B. Merkelbach, *Summa theologiae moralis* (Paris: Desclée, 1938), II, nr. 356; H. Davis, *Moral and Pastoral Theology*, II, p. 151; H. Jone-U. Adelman, *Moral Theology* (Westminster: Newman, 1947), p. 147.

[25] Ezechiel, 18:4.

[26] Genesis, 9:6.

[27] Numbers, 36:33.

[28] Romans, 13:4.

The problem of killing in war must also be considered at this point, as we study the term *innocent* as contained in the thesis. Soldiers who are conscripted or who have joined the army before a war, may usually presume that their country is in the right. If they are in insoluble doubt concerning the justice of their nation's cause, they should settle the doubt in favor of the government and obey. If the war is clearly unjust on the part of the soldier's own government, he may not engage in active combat. If a soldier wishes to join the army freely in time of war he must first satisfy himself that the war is just.[29] As regards the killing itself in a just war, there are two opinions. Some theologians maintain that the killing in a war may be only *indirect*. (The terms *direct killing* and *indirect killing* will be explained at the end of our definition of terms.) Dr. Cronin defends this view:

> War being of its nature an act of defense, it follows that killing in a war is indirect and not direct. It is never lawful to will directly a thing which is evil or unlawful or disallowed; but it is lawful *under certain conditions* to do an act, good or lawful in itself, for the sake of the good consequences which it produces, even though it is known that the same act will be attended by evil consequences also. In that case we are said to will those evil consequences indirectly only. . . . A nation goes to war in self-defense. For this, all that is necessary is to break down the resistance of the enemy, to put him out of action, and this, and what is necessary for this, a nation aims at directly. But death *as such* is not necessary for this, and therefore a nation may not aim directly at the death of the enemy. . . . Now in actual battle it would be ridiculous to expect a soldier to make this distinction and to use the instruments of war in such a way as to wound only and not to kill. But such precautions are possible in *devising* and supplying instruments of war.[30]

Many other theologians believe that the killing in a war is direct. Among these theologians are St. Thomas, St. Alphonsus, De Lugo,

---

[29] H. Davis, *Moral and Pastoral Theology*, II, p. 149; J. Aertnys-C. Damen, *Theologia moralis*, I, nr. 590; H. Noldin-A. Schmitt, *Summa theologiae moralis*, II, nr. 354; B. Merkelbach, *Summa theologiae moralis*, II, nr. 275-sub. 5; A. Koch-A. Preuss, *Handbook of Moral Theology*, II, p. 87.

[30] M. Cronin, *The Science of Ethics*, II, p. 666.

Laymann, Merkelbach, Damen, Noldin, and Davis.[31] This is the more common view. The soldiers of the nation with justice on its side are given the commission by a lawful authority to put to death the enemy for their objective aggression against peace, or because of their armed support of injustices.

> In a war obviously every soldier is the mandatory of his country. He is given the general authority to kill the enemy wherever he may find him, whether he be a soldier in actual battle, or one who aids and abets the cause behind the lines. This then includes not only all in arms but even the leaders of the enemy government who give the general war commands, or who have the authority to stop the war, yet do not do so. . . . If then direct killing in war may be justified, obviously, the natural law permits of such instruments of war which Father Cronin says are disallowed, such as poisoned or explosive bullets, the sole and necessary effect of which is to kill.[32]

It is to be remembered that an aggressive war is not necessarily an unjust war. The nation committing the first physical aggression may in reality be fighting a just war. The *status quo* against which the nation is fighting may be a grave evil that can be overcome in no other way.

Killing in Self-Defense

Under certain conditions one may kill a person who without authorization attacks him or another party. It is the more common opinion that the killing of an unjust aggressor may be only indirect,

---

[31] St. Thomas, *Summa theologica*, II. II, Q. 64. Art. 7; St. Alphonsus, *Theologia moralis*, Lib. III, Dub. V, Art. III (Vol. I, 409, Gaudé); J. De Lugo, *De justitia et jure* (Lyons: 1670), I, Sect. 4, nr. 107: P. Laymann, *Theologia moralis* (Venice: 1638), Lib. III, Tr. III, Pars. III, Cap. I. Assertio III; B. Merkelbach, *Summa theologiae moralis*, II, nr. 358; J. Aertnys-C. Damen, *Theologia moralis*, I, nr. 590; H. Noldin-A. Schmitt, *Summa theologiae moralis*, II, nr. 354; H. Davis, *Moral and Pastoral Theology*, II, p. 149.

[32] J. Sherman, "Aiming at Death," *The Ecclesiastical Review*, 108, I (Jan. 1943), p. 109.

in the manner that will be explained below.[33] No private individual has authority over the life of another person, and therefore to place by his own authority an act that has as its only immediate end the death of an unjust aggressor, is a grave crime. It must be remembered that even though a man makes an attack upon another, he does not forfeit his life to that other person. Accordingly, if a criminal attacks a man, the man may not aim at the criminal's death, but he may place whatever acts are necessary for protecting his own life, even if in doing this the criminal's death would follow as a concomitant effect. It would appear that in self-defense, the aim is always and first of all, the stopping of the attack.

> What is evident, therefore, is that in self-defense the only means chosen and the only means necessary is the stopping of the unjust aggression. The death of the aggressor is not aimed at and is not a necessary means of self-defense. When death occurs it is caused by us indirectly only and cannot be helped. From what precedes it is easy to gather the conditions necessary for a blameless defense. First the death of the aggressor must not itself be made an object of pleasure or be willed in itself; secondly, the defense must occur during or in the act of aggression (in ipso actu aggressionis) else it would be more than the stoppage of a charge; thirdly, not more violence should be used than is required to stop the attack; if more violence is used than is necessary, our act is more than one of defense, it is a new aggression; perhaps we may add a fourth condition for clearness sake—the act directed against us should be strictly one of aggression. I cannot kill a diseased man simply because of the danger, that, if he lives, I or others shall die.[84]

Lastly the thesis is concerned with *direct* killing. Direct killing means an action or omission which has no other immediate end

---

[33] M. Cronin, *The Science of Ethics*, II, p. 97; J. Aertnys-C. Damen, *Theologia moralis*, I, nr. 571; H. Noldin-A. Schmitt, *Summa theologiae moralis*, II, nr. 333; B. Merkelbach, *Summa theologicae moralis*, II, nr. 361-sub. 4; H. Jone-U. Adelman, *Moral Theology*, nr. 215, 3; A. Koch-A. Preuss, *Handbook of Moral Theology*, V, p. 125; St. Thomas, *Summa theologica*, II, II, Q. 64, Art. 7.

[84] M. Cronin, *The Science of Ethics*, II, p. 100. Cf. H. Davis, *Moral and Pastoral Theology*, II, p. 153.

than the death of a person. The death is intended as an end in itself or as a means to an end.[35] If a man is killed out of revenge, then the death is inflicted as an end desirable in itself and hence is direct killing. If a man is killed in order that a secret be maintained or that an end be put to his sufferings, then the death is inflicted as a means to an end, but this is also direct killing. In either of these cases the death inflicted is the only immediate effect. Direct killing must be understood in contradistinction to indirect killing, which may be defined as an action or omission having some other immediate effect in addition to the death of a person. Such a death, even when foreseen to follow an act, need not be intended in itself, but can be merely permitted.[36] Thus, if the brakes of an automobile fail and the machine begins to dash downhill into a large crowd, the driver may steer it aside even at the risk of running over one person. In this event there are two immediate effects—the avoidance of a great number of (at least probable) deaths and the (at least probable) death of one individual. Again, a lawful military objective may be bombarded in time of war even though as another immediate effect some civilians will lose their lives. These two examples illustrate indirect killing.[37]

PROOF OF THESIS: IT IS NEVER LAWFUL FOR MAN ON HIS OWN AUTHORITY TO KILL THE INNOCENT DIRECTLY.

At the outset it can be admitted that the thesis is not self evident to all men, at least in so far as it includes mercy killing. Many Christians who condemn suicide and murder, believe that euthanasia is consistent with Christian morality. This is evident from the number of ministers who have signed in favor of euthanasia.[38] Furthermore, it must be admitted candidly that without the support of the teaching Church, Revelation, and the general

---

[35] B. Merkelbach, *Summa theologiae moralis*, II, nr. 349; H. Noldin-A. Schmitt, *Summa theologiae moralis*, II, nr. 326; J. Aertnys-C. Damen, *Theologia moralis*, nr. 566; H. Davis, *Moral and Pastoral Theology*, II, p. 152.

[36] B. Merkelbach, *Summa theologiae moralis*, II, nr. 349; J. Aertnys-C. Damen, *Theologia moralis*, I, nr. 566.

[37] H. Jone-U. Adelman, *Moral Theology*, p. 146.

[38] Cf. supra Ch. 3.

heritage of the West, it is difficult to establish from reason alone that mercy killing, especially voluntary mercy killing, is always under every condition illicit. Though the argument from reason is objectively good, it requires more than reason to present a proof for the thesis that will carry conviction to all. In proving euthanasia illicit, this dissertation will appeal, therefore, not only to reason, but likewise to the great tradition of the West which has been Christian for nearly two thousand years. It will appeal also to the argument of the "wedge principle," and to man's natural desire to live.

Argument From Reason

This argument from reason presupposes belief in a personal God, the immortality of the human soul and the existence of a moral law binding all human beings. To the atheist, of course, this argument would have no appeal. However, since the vast majority of mankind does believe in God and in the other truths mentioned, the argument has a practical validity.

Supreme dominion over life belongs to God alone. God has sovereign dominion over all things, even over the essences of things. According to His providence He has given man a natural dominion over things of the earth inferior to himself to the extent that man may make use of them for his own utility.[39] Man has not full dominion over his life. He has only the use of it; and the natural law obliges man while using a thing which is under the dominion of another not to destroy it. The life of man is solely under the dominion of God. Wherefore, as man may not take his life, neither may another man take it (apart from exceptions given above), since another man would have even less dominion over it.[40]

An ample treatment of this dominion of God over life is contained in the works of Molina:

> Man is not the master of his own life and members as he
> is the master of money and of other external goods which

---

[39] J. De Lugo, *De justitia et jure* (Lyons: 1670), I, Sect. 4, nr. 102.
[40] *Ibid.*, p. 259, nr. 102 f.

pertain to him and which he possesses. The Lord indeed conceded to men dominion over external goods, and after the division of the things, each one disposes of those things which belong to him as of things of which he is truly the master; for that reason, by destroying them at his will, he does not sin against justice, since he destroys that which is his own . . . but dominion over life and members, the Author of Nature who created them, reserved to Himself, conceding only the use and administration of them to men.[41]

Molina proceeds to comment that should a man kill another innocent man or himself, he would not violate justice toward God in the strict sense because there cannot be strict justice in the relationship of man to God. Man would, however, commit a grave sin for it is even something higher than justice that binds man to God.[42] God is the Creator and hence, he remarks, it is not even fitting that God should give man the right to take his life. God gave the life to man and man did not bestow it on himself. Dominion over life goes with being the Creator of life.[43]

Laymann too sees in the taking of the life of an innocent person the violation of God's supreme dominion over life and in a sense the violation of justice to God.[44]

Similar views are expressed by the theologians, Lessius and Ballerini-Palmieri who repeat almost verbatim the argument of

---

[41] L. Molina, *De justitia et jure* (Venetiis: 1611), Vol. IV, Disp. I, nr. 1. "Homo non est dominus propriae vitae ac membrorum, sicut est dominus pecuniae et caeterorum bonorum externorum, quae ad ipsum spectant ac possidet. Dominium quippe bonorum externorum concessit Dominus hominibus et post rerum divisionem unusquisque de eis rebus, quae ad ipsum pertinent, disponit tamquam de rebus quarum vere est dominus; eaque de causa, destruendo eas pro libito, non peccat contra justitiam, quoniam destruit quod suum est . . . at vero dominium vitae et membrorum, auctor ipsius naturae qui ea contulit, sibi ipsi reservavit, solum usum et administrationem eorum hominibus concedens."

[42] *Ibid.*

[43] *Ibid.*

[44] P. Laymann, *Theologia moralis* (Venice: 1638), Lib. III, Tr. III, Pars. III, Cap. I, Assertio III.

Molina.[45] Billuart in commenting on the *Summa* expresses himself after the manner of the other theologians.[46]

De Lugo presents the best argument of all and it is the argument that is followed in this chapter.

> Now that man is not the master of life can be proved in this wise: because although man can receive dominion over things which are extrinsic to himself or which are distinct from him, he cannot, however, receive dominion over himself, because from the very concept and definition, a master is something relative, as a father or a teacher; whence as no one can be father or teacher to himself, so also he cannot be master of himself, for master always bespeaks some superiority with regard to that one of whom he is the master. Hence God Himself connot be Lord of Himself, although He possesses Himself most perfectly. Man, therefore, cannot be made the Lord of himself; he can, however, be the master of his operations, and therefore he can sell himself, and then it is not correct to say that he gives to another *simpliciter* dominion over himself, but only dominion over certain of his operations. . . . Therefore man can dispose of his own operations, of which he is lord but not of himself, or (which is the same thing), of his own life, dominion over which is not his and cannot be.[47]

---

[45] L. Lessius, *De justitia et jure* (Antwerp: 1617), Lib. II, Cap. 9, Dub. 6; A. Ballerini-D. Palmieri, *Opus theologicum morale* (Paris: 1899), Vol. II, Tr. VI, Sec. V, Cap. I, Dub. I, nr. 856.

[46] C. Billuart, *Summa Sancti Thomae hodierna academiarum moribus accomodata* (Paris: 1846), Vol. IV, Dissert. X, Art. 3.

[47] J. De Lugo, *De justitia et jure*, I, Disp. X, Sect. I, nr. 2. "Porro hominem non esse dominum suae vitae probari potest, quia licet homo potuerit accipere dominium aliarum rerum, quae sunt extra ipsum, vel quae ab ipso distinguuntur, non tamen potuit accipere dominium suipsius, quia ex ipso conceptu et definitione constat, dominus est aliquid relativum, sicut pater et magister; quare sicut nemo potest esse pater vel magister suipsius, ita nec potest esse suipsius dominus, nam dominus semper dicit superioritatem respectu illius cujus est dominus. Unde nec Deus ipse potest esse dominus suipsius, quamvis possideat perfectissime seipsum. Non potuit ergo homo fieri dominus suipsius; potest quidem esse dominus suarum operationum, et ideo potest vendere seipsum, et tunc dicitur improprie dari alteri dominium sui simpliciter, sed solum in ordine ad aliquas suas operationes. . . . Ergo homo potest disponere de suis operationibus quarum dominus est, non de seipso, vel (quod idem est) de vita sua cujus dominus non est, nec esse potest."

The argument from God's supreme dominion is likewise offered by Henno but in a very brief form.[48] Reiffenstuel's treatment is much like De Lugo's chief argument but it lacks the latter's precision.[49] Ferreres deals with the question in one sentence referring to man's dominion simply as a *dominium utile* and an indirect dominion while God alone has the *dominium plenum*.[50] In a word, it is the belief of Catholics that according to the natural law man does not have direct dominion over human life—either his own life, or the life of another man. Yet, euthanasia, whether it is voluntary euthanasia or compulsory euthanasia, is the destruction of human life. Destruction is an act proper to the master alone. Hence euthanasia violates God's absolute dominion over human life. The fact that the victim wills to die is an irrelevant consideration. The patient is innocent and hence apart from a divine command, may not be killed directly whether he wills it or not.

An objection might be presented, that the common good may sometime require the killing of an innocent person. If for example a diseased man became a grave danger to a community so that unless he were put out of the way many others would die, the state could kill the diseased patient for the common good. It might seem that the state would have the right to cut off a member of its body for the health and safety of the whole body, as a man may cut off a member of his body when it is necessary for the health of the whole body. This is confirmed by the fact that although the state does not have dominion over the life of the citizen; yet it is permitted to kill a citizen in punishment for a crime, because the punishment is useful and vital to the common good of the whole state. Therefore, when it is equally or even more necessary to kill an innocent member for the common good, it seems that it should be permissible.

Some might answer that for the health of the whole body a sick or evil member may be cut off, not a healthy one; thus in like

---

[48] F. Henno, *Tractatus moralis in Decalogi* Praecepta (Tornaci: 1711), De Quinto Praecepto, Art. I, Conclusio 1, nr. 749.

[49] A. Reiffenstuel, *Theologia moralis* (Munich: 1692), Tr. IX, Dist. 3, Quaest. 1, nr. 3.

[50] J. Ferreres, *Compendium theologiae moralis* (Barcinonae: 1925), Vol. I, Tr. V, Cap. I, nr. 487.

manner a malefactor may be cut off from the state, but not an innocent member.[51] The diseased man is an innocent member of the community and therefore to kill him, to administer compulsory euthanasia to him, is a violation of God's absolute dominion over life.

To such a defense it could be answered—and justly, we believe—that even a healthy member of the natural body of man may be cut off if it is necessary to save the life of the whole person, as for example when an arm is amputated in order that a person may escape a death trap. It might seem therefore that even a healthy member of society, an innocent member of the community, may be cut off for the sake of the whole. If such is permissible then the original thesis is unsound.

The true response to this objection, therefore, is to be found in the great difference between the members of the natural body of man and the members of the political body. A member of the natural body has no independent existence and no individual rights. The person itself has the right to use the members for they exist for the utility of the person. Wherefore properly they may be cut off when it is necessary for the conservation of the person, for the sake of which they exist.[52] It is a very different relation that exists between the individual human being and the state. The citizens do not exist for the sake of the state, but for their own sake, and to attain their own destiny.[53] A citizen does not serve the utility of the state in the sense that the state has full dominion over him. It is the state that exists for the utility of the citizen.[54]

> The origin and the primary scope of social life is the conservation, development and perfection of the human person, helping him to realize accurately the demands and values of religion and culture set by the Creator for everyman and for all mankind, both as a whole and in its natural ramifications. A social teaching or a social reconstruction program which denies or prescinds from this internal essential relation to God of everything that

---

[51] J. De Lugo, *De justitia et jure*, Vol. I, Sect. IV, p. 259, nr. 103.
[52] *Ibid.*, nr. 104.
[53] *Ibid.*
[54] *Ibid.*

regards man, is on a false course; and while it builds up
with one hand, it prepares with the other the materials
which sooner or later will undermine and destroy the
whole fabric. And when it disregards the respect due
to the human person and to the life which is proper to
that person, and gives no thought to it in its organization,
in legislative and executive activity, then instead of serving society, it harms it; instead of encouraging and
stimulating social thought, instead of realizing its hopes
and expectations, it strips it of all real value and re-
duces it to an utilitarian formula which is openly rejected
by constantly increasing groups.[55]

Hence the state may not cut off or kill an innocent citizen merely
for its utility, because this is an encroachment of a fundamental
personal right, at least when there is a question of compulsory
euthanasia. Even if a sick person were to agree to his own death
for the common good and undergo voluntary euthanasia, there
would still be committed a grave violation of God's absolute
dominion over life. Since man does not have full dominion over
his own life, he obviously cannot give up a dominion he does
not have.

There is another objection that may be advanced especially as
regards voluntary euthanasia. It might be maintained that life is
a gift and as such may be renounced. The answer, however, is that
life is not merely a gift. It is indeed a gift, but it has grave obligations inseparably affixed thereto both to God and to one's fellow
man. Life is given man not only for himself but also for the
service of his fellow man and God.

It may further be objected that even though God does have
absolute dominion over man's life, yet man does presume the
right to mutilate a member of that body when necessary for the
good of the whole. And furthermore, it seems, man may even
mutilate a part of his body, within certain limits, out of charity
for his neighbor.[56] If these two presumptions are permissible it
might seem likely that God would allow a man a merciful death

---

[55] Leo XIII, "Immortale Dei," ASS, Vol. 18 (1885), p. 161.

[56] B. Cunningham, *The Morality of Organic Transplantation* (Washington:
Catholic University Press, 1944), p. 100.

out of charity to himself and to those who must care for him. Once a presumption such as a minor or a major mutilation is allowed it would appear at least doubtful as to where to draw the line.

This objection is not valid. One can presume reasonably that when he has the administration of a thing belonging to another, a part may be sacrificed when it is necessary to save the whole. Hence it is a reasonable presumption that man may mutilate a part of his body when it is necessary to save the whole body. There is quite a difference from saying this and from saying that man may outrightly destroy what belongs to his master because to retain it causes pain and inconvenience. If a ship is carrying a precious cargo to its owner in a distant port, and it becomes evident to the ship's captain that a part of the cargo must be thrown overboard lest the whole cargo be lost, the captain may presume reasonably that the owner of the cargo would allow him to sacrifice a part of it in order that the whole cargo not be lost. But it would be unreasonable to conclude from this that therefore the captain could destroy the owner's whole cargo because some inconvenience is involved in taking care of it. The line of demarcation seems very evident. It must be remembered that in the two presumptions presented in the objection, it was actually for the prolongation of life that the mutilation was allowed.

Surely, it is illogical to argue that since man is permitted to take certain measures to prolong his life, he may therefore sometimes take measures to destroy it. Man has only the use of his life, and when man uses what belongs to another he has the obligation of taking ordinary care of it. When dealing with a human life, this ordinary care may involve some necessary mutilation for the preservation of that life. Hence man's right to presume a reasonable mutilation of his body not only does not argue against God's supreme dominion over life but rather confirms it. It is because God has the absolute dominion over life that man may even be required to mutilate his body, when this is a part of the ordinary care due what belongs to another.

Another objection that is often advanced by advocates of euthanasia is that to maintain that God would object to killing an incurable sufferer, a useless member of society, one who is a burden to all, is unreasonable. There is no point in forcing certain physically and mentally defective persons to live a miserable existence. To hold the contrary is to manifest a heartless attitude toward the sufferers themselves and to place an unbearable physical, emotional, and financial burden upon the families of many of these unfortunate persons.

The Christian has the only answer to this objection. He knows that no human being is a useless member either to himself or to society or at least need not be, no matter what his physical status may be. If the suffering patient is of sound mind and capable of making an act of divine resignation, then his sufferings become a great means of merit whereby he can gain reward for himself and also win great favors for the souls in Purgatory, perhaps even release them from their suffering.[57] Likewise the sufferer may give good example to his family and friends and teach them how to bear a heavy cross in a Christlike manner.

As regard those that must live in the same house with the incurable sufferer, they have a great opportunity to practice Christian charity. They can learn to see Christ in the sufferer and win the reward promised in the Beatitudes.[58] This opportunity for charity would hold true even when the sufferer is deprived of the use of reason. It may well be that the incurable sufferer in a particular case may be of greater value to society than when he was of some material value to himself and his community.

Argument From Western Tradition

The tradition of the West has been Christian for nearly two thousand years and it is interesting to note that never has this tradition sanctioned the direct killing of the innocent (apart from a divine command). There is no positive approval of direct killing of the innocent in any of the writings of the Fathers. Here there is at least an argument of silence against such a killing. If there had been an opinion in any sense common during those early days

---

[57] A. Tanquercy, *Synopsis theologiae dogmaticae* (Paris: Desclée et Socii, 1938), p. 797, nr. 1125; *Decretum de purgatorio* (DBU, 983, p. 342); II Machab., 12:43.

[58] St. Matthew, 5:7.

of Christianity to the effect that under certain circumstances man may kill the innocent directly, this opinion would have found mention in some of those ancient writings. That such is not the case, is in itself an implicit disapproval of direct killing. On the contrary, Christianity from the very beginning opposed self-destruction and homicide and this is positively indicated in the writings of some of the Fathers and Doctors of the Church. St. Augustine affirms this common tradition of the Christian West:

> The commandment is, 'Thou shalt not kill man,' therefore neither another nor yourself, for he who kills himself still kills nothing else than man.[59]

The value of this traditional argument toward proving the thesis was noted by St. Augustine himself.

> It is not without significance that in no passage of the holy canonical books can there be found either divine precept or permission to take away our life, whether for the sake of entering on the enjoyment of immortality, or for shunning, or ridding ourselves of any evil whatever.[60]

· The Western tradition against direct killing of the innocent found earlier expression in the writings of Lactantius, an Ante-Nicene Father.

> For if a homicide is guilty because he is a destroyer of man, he who puts himself to death is under the same guilt, because he puts to death a man. Yea, that crime may be considered to be greater, the punishment of which belongs to God alone. For as we did not come into this life of our own accord; so, on the other hand, we can only withdraw from this habitation of the body which has been appointed for us to keep, by the com-

---

[59] St. Augustine, *De civitate Dei*, Lib. I, 20 (CSEL, 40, p. 38). ":restat ut de homine intellegamus, quod dictum est, non occides; nec alterum ergo, nec te, neque enim qui se occidit, aliud quam hominem occidit."

[60] St. Augustine, *De civitate Dei*, Lib. I, 20 (CSEL, 40, p. 38). "Neque enim frustra in sanctis canonicis libris nusquam nobis divinitus praeceptum permissumve reperiri potest, ut vel ipsius adispicendae immortalitatis vel ullius carendi cavendive mali causa, nobisinctipsis necem inferamus."

mand of Him who placed us in this body that we may
inhabit it, until He orders us to depart from it, and if any
violence is offered to us, we must endure it with
equanimity, since the death of an innocent person cannot
be unavenged, and since we have a great Judge who
alone always has the power of taking vengeance in His
hands.[61]

Many of the other Fathers express themselves in like manner,
especially St. Cyprian,[62] St. Ambrose,[63] St. Irenaeus,[64] St. Athanasius,[65] St. Bede the Venerable,[66] and St. Anselm.[67] As a matter
of fact there are seventy-seven Saints, Fathers or Popes listed in
Migne who in some way indicate in their writings the Western
tradition against the direct killing of the innocent.[68] Among this list
are such teachers as St. Benedict, St. Fulgentius, Pope Innocent
III, Pope St. Gregory the Great, Pope St. Leo the Great, and Peter
the Lombard. St. Thomas, too, stands sternly against all forms of
self-destruction or murder.

> . . . the life of the righteous men preserves and forwards
> the common good, since they are the chief part of the
> community. Therefore it is in no way lawful to slay the
> innocent.[69]

---

[61] Lactantius, *Divinarum Institutionum*, Lib. III, 18 (CSEL 19, p. 237).
"Nam si homicida nefarius est, quia hominis extinctor est, eidem sceleri
obstrictus est, qui se necat, quia hominem necat. Imo vero maius esse id
facinus existimandum est, cuius ultio soli Deo subiacet. Nam sicut hoc
domicilio corporis, quod tuendum nobis adsignatum est, eiusdem iussu re-
cedendum est, qui nos in hoc corpus induxit, tamdiu habituros, donec iubeat
emitti, et si vis aliqua inferatur, aequa mente patrandum, cum extincta in-
nocentis anima inulta esse non possit, habeamusque iudicem magnum, cui
soli vindicta in intergro semper est."

[62] St. Cyprian, *De charitate inter fratres* (CSEL, 4, p. 737).

[63] St. Ambrose, *In Psalmum enarratio* XXXVI (PL. 14, p. 975).

[64] St. Irenaeus, *Adver. Hacreses* XXIII (PG. 18, p. 729).

[65] St. Athanasius, *De operibus charitatis* (PG. 18, p. 880).

[66] St. Bede, *De scandalo* (PL. 94, p. 442).

[67] St. Anselm, *De pace et concordia* (PL. 158, p. 1015).

[68] Cf. *Quintum Praeceptum* (PL. 220, p. 394).

[69] St. Thomas, *Summa theologica*, II, II, Q. 64, Art. 6. ":vita autem
justorum est conservativa, et promotiva boni communis: quia ipsi sunt
principalior pars multitudinis; et ideo nullo modo licet occidere innocentem."

In the Middle Ages one scarcely hears of suicides, and one would be tempted to deny their occurrence during those times, were it not for the numerous acts of ecclesiastical and civil legislation condemning and declaring punishable even the very attempt to commit suicide.[70]

The Church had legislated against self-destruction as early as the year 533 at the Council of Orleans. It was determined at that time not to accept the offerings of a man who died at his own hands.[71] Later synods, for instance that of Braga in 563,[72] and Auxere in 576,[73] renewed the penalties against the offenders. In the ninth century the Council of Troyes warns against a free-will death.[74] Even the Catechism of the Council of Trent, published at the command of Pope Pius V, reminds the faithful that murder and suicide are forbidden by Divine Command.

> With regard to the person killed, the law extends to all. There is no individual, however humble or lowly his condition, whose life is not shielded by this law.
> It also forbids suicide. No man possesses such power over his own life as to be at liberty to put himself to death. Hence we find that the Commandment does not say: *Thou shalt not kill another,* but simply: *Thou shalt not kill.*[75]

The tradition was so strongly set against suicide in the Middle Ages that not only was Christian burial denied the victim but in addition to this his goods and property were confiscated by the civil authority. Being excluded from the Christian cemetery,

---

[70] A. Frenay, *The Suicide Problem in the United States* (Boston: Gorham, 1926), p. 66.

[71] J. Mansi, *Sacrorum Concilliorum Nova et Amplissima Collectio* (Arnheim: 1901-1927), VIII, p. 837.

[72] J. Mansi, *Sacrorum Concilliorum Nova et Amplissima Collectio*, IX, p. 779.

[73] J. Mansi, *Sacrorum Concilliorum Nova et Amplissima Collectio*, IX, p. 833.

[74] J. Mansi, *Sacrorum Concilliorum Nova et Amplissima Collectio*, XIV, p. 932.

[75] J. McHugh-C. Callan, *Catechism of the Council of Trent* (New York: Wagner, 1943), p. 423.

the suicide received an ignominious burial on the highway with a stake driven through his body.[76] This severity of civil law was in force throughout the Middle Ages. There was no exception made even for incurable sufferers. And surely there must have been some suicides in this class.

These penalties were gradually mitigated and in the year 824 the English Parliament permitted a suicide's burial in the church-yard but only between nine and twelve at night.[77]

This past tradition of the Christian West indicates how strongly the people felt against the direct killing of the innocent and especially against suicide.

When the Church laws were codified in 1918 this Western culture was still evidenced, for a definite penalty was placed on suicide. Persons guilty of deliberate suicide are to be denied ecclesiastical burial unless they have before death given some signs of repentance.[78] The same penalty is to be placed on those dying from wounds received in a duel.[79] This is in punishment for exposing their lives and the lives of their adversaries to grave danger. Again, the stand of the Church against the direct killing of unborn infants, the practice of abortion, is another indication of her condemnation of the direct killing of the innocent. The Code states that persons who procure abortion, the mother not excepted, automatically incur excommunication reserved to the Ordinary at the moment the crime takes place.[80] The most *ad rem* decree against a direct killing of the innocent was issued by the Holy Office in 1940 and deals expressly with state eugenic murder or in other words compulsory euthanasia.

> Q: Is it lawful upon the mandate of authority directly to kill those who, although not having committed any crimes deserving of death, are however, because of the psychic or physical defects, unable to be useful to

---

[76] A. Frenay, *The Suicide Problem in the United States*, p. 66.

[77] *Ibid.*

[78] *Codex Iuris Canonici*, Can. 1240.

[79] *Ibid.*

[80] *Ibid.*, Can. 2350.

the nation but, rather, are considered a burden to its vigor and strength?

A: No, because it is contrary to the natural and the divine positive law.[51]

Then there is also a passage from Pope Pius XII in his Encyclical Letter *Mystici Corporis* that condemns euthanasia.

Conscious of the obligations of Our high office, We deem it necessary to reiterate this grave statement today, when to Our profound grief We see the bodily-deformed, the insane and those suffering from hereditary disease, at times deprived of their lives, as though they were a useless burden to society. And this procedure is hailed by some as a new discovery of human progress, and as something that is altogether justified by the common good. Yet what sane man does not recognize that this not only violates the natural and Divine law written in the heart of every man, but flies in the face of every sensibility of civilized humanity? The blood of these victims. all the dearer to Our Redeemer because deserving of greater pity, "cries to God from the earth."[52]

Even a more recent statement was made by the Holy Father, Pope Pius XII, in which he condemns abortion and euthanasia. The statement was made in a talk given in Italy in May of 1948.[53]

This tradition of the West is based to a great extent on Sacred Scripture. The Scriptures viewed only as a historical record of the ethics of a people will indicate the tradition of the people. In the case at hand the Scriptures indicate a tradition opposed to the direct killing of the innocent. It should be remembered too, that as regards these historical books of Sacred Scripture, even the

---

[51] AAS, Vol. XXXII (1940), p. 553. "Q: Num licitum sit, ex mandato auctoritatis publicae, directe occidere eos qui, quamvis nullum crimen morte dignum commiserint, tamen ob defectus psychicos vel physicos nationi prodesse jam non valent eamque potius gravare ejus vigori ac robori obstare consentur?

A: Negative, cum sit jure naturali ac divino positivo contrarium."

[52] Pius XII, "Mystici Corporis," AAS, Vol. 35 (1943), p. 239.

[53] Pius XII, "Integrity of the Human Body," A radio talk delivered May 25, 1948. Cf. N.C.W.C. *News Service Letter*, 5/24/48.

pre-Christian culture is recorded. Therefore, as far as the Old Testament is concerned the Scriptures record a tradition of over four thousand years. This is of special importance for it is in the Old Testament writings that we find the most explicit statements condemning direct killing of the innocent. Hence we can conclude that not only the Western tradition of the past two thousand years, but even its pre-Christian culture, condemned direct killings apart from divine command.

The innocent and the just person thou shalt not put to death.[84]

The innocent and the just thou shalt not kill.[85]

It might also be mentioned as a negative argument that in the entire Old Testament there are only seven instances of suicide. The most commonly known instances are those of King Saul and of Achitophel, the councillor of Absolom. The other suicides are: Eleazar, Razias, Zambri, Abmimeleck and Ptolemee.[86] In the New Testament there is the case of Judas who hanged himself. Otherwise it appears that suicide was a rare occurrence among Jews, at least up to the year 70 A.D.

The Western tradition is likewise evidenced in the writings of the theologians of the Roman Church. For the most part the West took its ethics from the teaching of these men. Many of these were mentioned under the Fathers and Doctors of the Church. Here a few of the past and the present teachers will be mentioned. The important point is, however, that all of these theologians teach that it is never lawful for man on his own authority to kill the innocent directly. This has been stated by such theologians as Molina,[87] De Lugo,[88] Lessius,[89] Laymann,[90] Billuart,[91] Sylvius,[92]

---

[84] Exodus, 23:7.

[85] Daniel, 13:53.

[86] II Kings, 30:4; II Kings, 17:23; II Mach., 6:27; II Mach., 14:41; III Kings, 16:18; Judges, 9:54; I Mach., 11:18.

[87] L. Molina, *De justitia et jure,* IV, Disp. I, nr. 1.

[88] J. De Lugo, *De justitia et jure,* I, Disp. X, Sect. I, nr. 2.

[89] L. Lessius, *De justitia et jure,* IV, Iib. II, Cap. 9, Dub. 6.

[90] P. Laymann, *Theologia moralis,* Lib. III, Tr. III, Pars III, Cap. 1.

[91] C. Billuart, *Summa Sancti Thomas hodierna academiarum moribus accomodata,* IV, Dissert. X, Art. 3.

[92] F. Sylvius, *Commentarii in totum secundam secundae S. Thomae Aquinatis* (Venice: 1726), Vol. III, Q. 64, Art. 5.

and the more modern theologians such as Noldin-Schmitt,[93] Aertnys-Damen, and Merkelbach.[94]

The tradition of the West is therefore sternly set against any form of direct killing of the innocent. It is true that the Western tradition does not explicitly condemn merciful euthanasia, because there was no need to do so since no one conceived the idea under a Christian civilization to kill the suffering in order to relieve them of their suffering. But by condemning all forms of direct killing of the innocent the tradition does implicitly condemn euthanasia, for euthanasia, whether voluntary euthanasia or compulsory euthanasia, is direct killing of the innocent. A patient is an innocent person. The fact that the patient is incurably ill, wills euthanasia, and is useless to the community, does not in any way change the fact of the person's innocence. For man to take the life of an innocent person directly, even his own life, for any reason whatsoever, apart from a divine command, has always been against the conscience of the West.

Euthanasia is a reversal from culture to barbarism. Even in most of the savage tribes that were studied in chapter two of this dissertation, there were no records of mercy killing in the strict sense. Most of those cases were referred to as quasi-euthanasia and represented abandonment of the aged when in reality little else could be done under their primitive living conditions. Only in the most barbaric tribes was there such direct killing of the aged and the sick.[95] The advocates of euthanasia must face the fact that they are opposing the traditional Christian teaching on the value of a human life. They must go back beyond two thousand years to a world before Christ, or to some pagan land that has never known Christ, and even then seek a most barbaric tribe to find a practice like to mercy killing.

Argument From the "Wedge Principle"

The "wedge principle" means that an act which, if raised to a general line of conduct would injure humanity, is wrong even in

---

[92] H. Noldin-A. Schmitt, *Summa theologiae moralis,* II, nr. 338.

[94] J. Aertnys-C. Damen, *Theologia moralis,* I, nr. 573; B. Merkelbach, *Summa theologiae moralis,* II, nr. 350.

[95] Cf. *supra,* Ch. 2, p. ??

an individual case. Ordinarily the act even individually is evil, but it can happen in exceptional instances that the act causes no harm but nevertheless on account of the common danger there is a general prohibition. By this principle one may conclude that any course of conduct that works destruction when practiced generally may not be permitted in an individual case. Thus divorce and remarriage might be harmless in some individual case. It may be that the innocent party and her children would be greatly benefited by her second marriage. Perhaps in this particular case there may be many good reasons why a second marriage would be helpful to all concerned, especially the children. Nevertheless, if one exception is allowed to the rule of no marriage after a divorce, other exceptions will soon follow and society will suffer much. (We are concerned with the natural law—not with any exceptions which God Himself may have introduced.)

This principle of the wedge may be applied to euthanasia, both voluntary euthanasia and compulsory euthanasia. Here for the sake of argument it will be presumed that the suffering patient wishes euthanasia and that no evil effects will result to his friends or the common good from the single act of administering the euthanasia to him. Nevertheless, euthanasia must not be administered, for to permit in a single instance the direct killing of an innocent person, would be to admit a most dangerous wedge that might eventually put all life in a precarious condition. Once a man is permitted on his own authority to kill an innocent person directly, there is no way of stopping the advancement of that wedge. There exists no longer any rational grounds for saying that the wedge can advance so far and no farther. Once the exception has been admitted it is too late; hence the grave reason why no exception may be allowed. That is why euthanasia under any circumstances must be condemned. We are making use of this as a secondary argument; for the primary argument is found in the intrinsic malice of the direct killing of an innocent person. But even one who would not admit this should acknowledge the value of the present argument.

If voluntary euthanasia were legalized, there is good reason to believe that at a later date another bill for compulsory euthanasia would be legalized.[96] Once the respect for human life is so low that an innocent person may be killed directly even at his own request, compulsory euthanasia will necessarily be very near. This could lead easily to killing all incurable charity patients, the aged who are a public care, wounded soldiers, captured enemy soldiers, all deformed children, the mentally afflicted, etc. Before long the danger would be at the door of every citizen.

Politics also would enter and perhaps political parties would use the law to destroy *personas non gratas*. This might be done by getting a court order stating that the person in question is an incurable mental case. The possibilities of abusing euthanasia are many. Dishonesty and graft have found their way into nearly every field. There would be many opportunities for these evils to operate under legalized euthanasia to the great danger of society. It is common knowledge that not infrequently a person goes into court to seek the administration of property by having the real owner declared insane. A more effective way of securing property would be open to unscrupulous relatives once euthanasia is legalized.

The most outstanding examples of the extremes to which legalized euthanasia can go are found in the mass eugenic murders within Germany and her conquered territories during the war. Those considered by the state as physically unfit or mentally unfit could be put to death as they were regarded as a grave burden to the common good. Likewise any undesirable citizen could be exterminated, whether man, woman or child. As a rule the death was an easy death in a gas chamber.[97] Many present day advocates of euthanasia might accept this procedure as within reason since the unfit and the undesirable are no benefit to society. The point is, however, that all persons unfriendly to the Nazi Government were considered by those in power as undesirable citizens. Those who held a philosophy contrary to that of the government might well be considered as mental cases. Hence many thousands were put to death without mercy.[98] It is estimated that thousands of

[96] J. Brown, "Taking Life Legally," *Magazine Digest* (March 1937), p. 43.

[97] A. Kinkel, *Put to Death Legally* (London: Davis Co., 1946), p. 37.

[98] *Ibid.*, p. 39.

Jews in Austria alone were sent to their death by the government for no reason save that they were non-arian and undesirable to the state. Reports indicate that these murders were considered legal by reason of existing compulsory euthanasia legislation.[99] At the time these laws were enacted it was thought that only the incurable mental cases, monstrosities, and the incurables that were a burden to the state, would be put to death. However, once the state held this power of life and death over even the innocent members of society, the lives of all the citizens were in danger.

This example of the Nazi government in Germany with regard to euthanasia might be followed by any other government once that government possessed such a power. One might say that the mentality of the American people would never accept such abuse of human life. The point is, however, that once the American people depart so far from Christian tradition as to allow an innocent person to be put to death legally because he is a monstrosity or wills euthanasia, it is difficult to predict how much further this abuse of the power over human life would go.

It is obvious, therefore, that if euthanasia is legalized, even with strict limitation, it would lead to many downright murders and hence by reason of the "wedge principle" may not be allowed in any individual case.

It is interesting to note that St. Augustine used the argument of the "wedge principle" in proving that a lie is never permissible. He maintained that if exceptions were allowed, more exceptions would be invented to a staggering and harmful degree.[100] It is important to note, however, that we do not hold that euthanasia is bad merely because if raised to a general line of conduct it would cause great harm. We hold that it is bad in itself.

Argument From Man's Desire to Live

Every being has an inclination, or appetite, to fill up the measure of its adequate perfection, and all which is in any way capable of satisfying such inclination is said to be good for that being.[101] All

---

[99] *Ibid.*, p. 40.

[100] St. Augustine, *Contra mendacium* (PL. 40, 544).

[101] M. Shallo, *Scholastic Philosophy* (Philadelphia: Hirschfeld Bros., 1918), 134.

creatures, even the plants and inanimate substances, have a natural tendency or affinity implanted in them, which impels them blindly toward what is "suitable to and perfective of their nature, independently of all cognition on their part."[102] So much does every appetite tend toward good that St. Thomas defines good as the object or end of appetite.[103]

Man, like the rest of creation, tends naturally toward the continuance and perfection of his being. This is true of every power in man—vegetative, sensuous, and intellectual.

> The will from its very nature aims at happiness in the attainment of some end, and therefore, also at the well being and development of the individual. The sensuous appetites, like that for food, aim at the fuller life and development of man on his sensuous side. Such vegetative tendencies or appetites as growth and the digestive movements are directed to the well-being or betterment of the substance of the body. It is impossible that any appetite set up in us by nature should be directed to any other thing than the fuller being of the individual. It is impossible that it should aim at nothingness or at destruction.[104]

St. Thomas uses this argument concerning the natural appetite in man to prove suicide unnatural, for he says:

> It is altogether unlawful to kill oneself . . . because everything naturally loves itself, the result being that everything naturally keeps itself in being, and resists corruption so far as it can. Wherefore suicide is contrary to the inclination of nature. . . .[105]

This inclination in man to prolong his life is verified in practice. It must be remembered that man differs from all other earthly crea-

---

[102] M. Shallo, *Scholastic Philosophy*, p. 236.

[103] St. Thomas, *Summa theologica*, I, Q. 6, Art. 2.

[104] M. Cronin, *The Science of Ethics*, II, p. 52.

[105] St. Thomas, *Summa theologica*, II, II, Q. 64, Art. 5. "seipsum occidere est omnino illicitum . . . quia naturaliter quaelibet res seipsam amat; et ad hoc pertinet, quod quaelibet res naturaliter conservat se in esse, et corrumpentibus resistit, quantum potest: et ideo quod aliquis seipsum occidat, est contra inclinationem naturalem. . . ."

tion in that he possesses an intellect and a will. Though his vegetative and sensuous powers tend necessarily to his continued existence, yet his will is free either to work toward the common goal of a healthy life, or to choose self-destruction. The fact that society exists today indicates that man throughout the ages has chosen to live. Man's desire to live is indicated by the great care given human life, from pre-natal care to the care given the aged in the social institutions throughout the world. Science has as its greatest goal the prolongation of life. The many health campaigns carried on within the United States, together with the drives to overcome cancer, tuberculosis and heart disease, are all very indicative of man's desire to live. Even the daily care man gives his physical and mental health testifies to this natural desire to live. War itself is an argument for the value of life, for man's greatest fear in war time is that either he or some dear friend may be killed. All of these facts make most obvious that universally man loves life and wants to live.

From this universal desire to live we have a strong argument against the direct killing of the innocent, and hence against all forms of euthanasia.

An objection might be raised that all healthy men desire to live but not all incurable sufferers. The very fact that euthanasia is so popular today is an indication that some incurable sufferers desire to die. Therefore the basic principle is false.

The objection is not valid. The incurable sufferer desires in the first place not to die but rather to recover or at least to be free from pain. Those who seek euthanasia do so because of the great pain. If this pain would cease they would rather live, even though they would not recover. The incurable sufferer's basic desire like the desire of all men is *to live* a healthy life. It is this basic natural desire to which the principle refers. The fact that the incurable patient may wish to end his sufferings, even by euthanasia if necessary, does not argue against his basic natural desire to live, and certainly it does not argue against the universal desire of man to live. It only indicates that man can have a confused sense of values and choose an evil under some aspect of good. Even in this case the incurable sufferer is violating a natural appetite for continued life.

POSSIBILITY OF INVINCIBLE IGNORANCE

It is likely that many of the present day advocates of mercy killing are in good faith. They may see that homicide is illicit and that suicide is illicit but they cannot see that euthanasia is necessarily always sinful. They will admit that human life is sacred, they may have respect for the fifth Commandment but somehow they feel that voluntary euthanasia is within the limits of morality.

Theologians admit that man may be invincibly ignorant of the more remote conclusions from the natural law. The distance removed from first principles increases this possibility. When man is occupied chiefly with material things he loses a true sense of values and hence his judgments are often incorrect.[106] Likewise man is dependent to a great extent on the culture about him and if this culture is pagan it is likely that he will be in good faith about many violations of the more remote conclusions of the natural law.

This is no argument against the universality of the natural law. All men do participate in the natural law. Monsignor Cooper makes this fact evident.

> The people of the world, however much they may differ as to the details of morality, hold universally, or with practical universality, to at least the following basic precepts. Respect the Supreme Being or the benevolent being or beings who take his place. Do not "blaspheme." Care for your children. Malicious murder or maiming, stealing, deliberate slander or "black" lying, when committed against a friend or unoffending fellow clansman or tribesman, are reprehensible. . . .[107]

The point is that not all are aware of the natural law to the same degree. All have inherent inclinations and at least some power of reason, so we believe that all are cognizant of the natural law at least *in actu primo*. Beyond that it varies with different

---

[106] S. Bertke, *The Possibility of Invincible Ignorance of the Natural Law* (Washington: Catholic University Press, 1941), p. 72.

[107] J. Cooper, "The Relations between Religion and Morality in Primitive Culture," *Primitive Man,* Vol. IV, No. 3, pp. 33 ff.

people. We present here a division of the natural law from the standpoint of those who recognize it:

A) Most Universal Principles: Do good, avoid evil, act according to human nature. (No one with the use of reason can be ignorant of these.)
B) Conclusions Immediately Deduced: The ten Commandments (with the exception of the determination of the Sabbath as the Lord's Day), at least as generally understood under ordinary circumstances. *Per se,* people are not inculpably ignorant of these. *Per accidens,* perhaps some can be ignorant of these for a time.
C) Remote Conclusions: Known in a greater or less measure, depending on education, culture, etc. Of these man can easily be ignorant.[108]

It is our opinion that euthanasia would come under this third division and hence many men, even good Christians, could be inculpably ignorant of the fact that euthanasia is against the natural law and natural ethics.

Among the evident deductions from the first principle of the natural law is the precept. *Thou shalt not kill.* Proof that it is an easy deduction may be had by both the *a priori* and *a posteriori* methods. By the former method it will be seen that the precept is but a combination of two first principles both of which are norms guiding fundamental human inclinations. To arrive at the conclusion, *Thou shalt not kill* is but to apply the self-evident principle, *Do unto others as you would have them do unto you,* to another self-evident principle guiding man to the conservation of his being. The former principle flows from the social nature of man, while the latter is based upon a tendency man has in common with other things. By thus reasoning from self-evident principles based on the very nature of man it is seen that the precept expressed by the Fifth Commandment of the Decalogue will be an evident deduction.

---

[108] B. Merkelbach, *Summa theologiae moralis,* I, nr. 251.

The *a posteriori* argument has been seen previously. No known tribe, primitive or otherwise, has condoned indiscriminate killing. Though primitive people have often qualified the general principle, *Thou shalt not kill,* by a distinction between their own tribes and other peoples, this universal condemnation of indiscriminate killing is a powerful argument for the conclusion already made from the nature of man, i.e., that the Fifth Commandment is an evident deduction from first principles.

This suggests the query whether people can be invincibly ignorant of the evil of euthanasia—the direct killing of a man in order to eliminate suffering or for the apparent good of the state.

In the theoretical treatment of the possibility of invincible ignorance the conclusion was made that *per se* such ignorance is not to be admitted concerning proximate and evident conclusions from first principles. However, it was conceded that *per accidens* the subject may conceive an action as justifiable in practical action surrounded with all its circumstances while fully admitting the general prohibition. This would hold in the present consideration.[109]

---

[109] S. Bertke, *The Possibility of Invincible Ignorance of the Natural Law,* p. 101.

CHAPTER V

SPECIAL PROBLEMS

Therapeutic Euthanasia

This euthanasia has been defined as: "that procedure whereby a physician eases the pains of the dying by the use of therapeutic doses of narcotics."[1] Therapeutic euthanasia is *non-lethal* and in itself lawful; however, certain moral problems are involved.

A distinction must be made between death's physical agony, and death's mental agony. For the former it is permissible to use sedatives, when the pain is very severe, even to the point of coma if the patient is spiritually prepared to die, but for the latter (mental agony) it is not permissible to use sedatives to this extent, unless the mental agony is so severe as to endanger the patient's act of spiritual resignation to the Will of God. In general it is not lawful to render a person unconscious by drugs just because he is in fear of approaching death, since this is a part of the sacrifice that God demands for the sins and the faults of life; and even when these have been atoned for, there is great spiritual value in accepting death with love and devotion to God. However, to mitigate pain coming from the illness that is causing death is not considered morally wrong, if the person is properly disposed.[2] Under no conditions may therapeutic euthanasia be administered to a patient against the patient's will.[3]

Confusion may result from a wrong understanding of the expression, death agony. In a popular sense it means simply an ex-

---

[1] Cf. *supra*, Ch. I, p. ?; also p. ?.

[2] S. Woywod, "The Use of Morphine and Other Opiates in Death Agony," *Homiletic and Pastoral Review*, XXXVII (1937), p. 1199.

[3] H. Noldin-A. Schmitt, *Summa theologiae moralis* (New York: Pustet, 1940), I, p. 343, nr. 349.

treme and prolonged period of physical suffering.[4] But in the technical sense the death agony is defined as: "the struggle, *frequently unconscious,* which often precedes death."[5] Physical suffering is not, therefore, essential to the death agony. It is obvious that that which is frequently unconscious is necessarily frequently painless. It is therefore likely that many undergo a death agony without any physical pain. Concerning this matter it is interesting to note that the *Rituale* does not imply a painful struggle during the death agony. The idea expressed is rather that a *contest* is in progress."

Obligation to Prolong the Life of Incurables

Theologians teach that *per se* one must use ordinary means to conserve the lives of incurable sufferers but not necessarily extraordinary means. They further maintain that the common estimation of men will indicate what is and what is not an ordinary means.[7] In many cases the distinction between ordinary means and extraordinary means will depend on the cost involved, or the gravity of the pain it will entail. A sick person in a grave illness but with hope of recovery, sins by not sending for a physician and taking the prescribed medicine, if this can be done easily. On the other hand, no one, not even a very wealthy person is obliged, *per se,* to call in a very expensive physician, or to accept a remedy that involves grave pain for a prolonged period.[8] There is an absolute norm beyond which means are *per se* extraordinary. When operations are considered, again relativity enters into the judgment of what is an ordinary means of prolonging life. As science advances, the extraordinary means of a decade ago become the ordinary means of today. It would seem today that even the amputation of an arm or

---

[4] *A New English Dictionary on Historical Principles* (Oxford: Clarendon Press, 1897), I, p. 188.

[5] *The Century Dictionary and Cyclopedia* (New York: Century Co., 1904), I, p. 115.

[6] H. Henry, "The Death Agony and Euthanasia," *Homiletic & Pastoral Review,* XXXVI (Nov. 1935), p. 123; A. Koch-A. Preuss, *Handbook of Moral Theology* (3. ed., rev., St. Louis: Herder, 1926), III, 91; H. Noldin-A. Schmitt, *Summa theologiae moralis,* I, p. 343, nr. 349.

[7] H. Noldin-A. Schmitt, *Summa theologiae moralis,* II, p. 308, nr. 325.

[8] *Ibid.*

leg would be an ordinary means of prolonging life.[9] In times past, such an operation was considered by theologians as an extraordinary means.[10] However, even today the amputation of both arms and both legs would seem to be an extraordinary means of prolonging life.

There is a great amount of relativity involved, therefore, in determining the ordinary and extraordinary means of prolonging a life. Ordinary means and extraordinary means are relative frequently to such circumstances as, time, place, cost, fear, and the advances of science. It is our opinion that the terms, "ordinary means" of preserving life and "extraordinary means" are relative also to the patient's physical condition. An aged woman sick unto death with cancer would not have to use the same means toward prolonging life as a young girl ill for the first time in her life with a hopeful future ahead. For this aged woman an operation which might prolong her life a few months or a year would be an extraordinary means. For the young girl even a more serious operation might be an ordinary means. The following two principles will help us determine in an individual case where to draw the line between ordinary and extraordinary means of prolonging life. *A natural means of prolonging life is, per se, an ordinary means of prolonging life, yet per accidens it may be extraordinary.* The second principle affirms: *An artificial means of prolonging life may be an ordinary means or an extraordinary means relative to the physical condition of the patient.* (Of course, it is relative to other matters also such as time, place, financial conditions, etc., but at present the interest is in establishing the fact that it is relative to the physical condition of the patient.)

Among the natural means of preserving life would be included such means as proper clothing, housing, physical recreation, good food, regularity at meals, etc. As artificial means, we may understand such means as major and minor operations, x-ray treatments, blood transfusions, intravenous feeding, radium treatments, psychotherapy, oxygen tents, iron lungs, all germicides and antiseptics, and even the taking of prepared medicines.

---

[9] *Ibid.*
[10] *Ibid.*

PRACTICAL APPLICATIONS

The following examples are offered to illustrate the moral principles set forth in chapter four. These examples concern either euthanasia in its strict sense of mercy killing, or problems closely related to it.

(A) A respected citizen kills his wife who had been bedridden for several years as a result of an incurable illness. He killed his wife at her own wish.

His act is, of course, gravely sinful. By killing his wife he usurped God's supreme dominion over human life.[11] He also sinned against the virtue of charity, for by killing his wife he deprived his wife of an objective good, her life.[12] The fact that his wife willed euthanasia does not change the situation.

(B) A physician, knowing that his patient is in grave pain and will never recover, gives her an overdose of sleeping pills. The next day the patient dies while asleep.

This is a case of compulsory euthanasia for the patient had not asked for release and may be presumed to will a natural death. The physician violated not only the virtue of charity but also the virtue of justice.[13]

(C) A Catholic girl, a nurse in a public institution, is called upon to prepare a victim of osteomyelitis for voluntary euthanasia. (We presume in this case that euthanasia is legal before civil law.) The preparation consists in the transfer of the patient from the sick bed to the cot and then wheeling the patient into the lethal gas chamber. The nurse fears that if she refuses she may lose her position and hence follows the orders given.

The nurse gives proximate cooperation toward a sin of murder. Even granting that her cooperation was only material, yet material cooperation is illicit unless there be a grave reason for placing the action, and indeed a proportionately grave reason.[14] In this case

---

[11] J. De Lugo, *De iustitia et iure* (Leipzig: Sumpti Laurentii, 1670), I, p. 259, nr. 102.

[12] B. Merkelbach, *Summa theologiae moralis* (Paris: Desclee, 1938), II, p. 360, nr. 359.

[13] *Ibid.*, V, p. 209.

[14] B. Merkelbach, *Summa theologiae moralis*, I, p. 400, nr. 489.

the material cooperation is very proximate, and the evil of euthanasia itself so great, that even the fear of losing her position would not seem to justify her following the orders.

(D) A Catholic lawyer is asked by his client to help him prepare the legal request for voluntary euthanasia. (We are supposing again that euthanasia is legal.) The lawyer accepts and draws up the necessary papers to be submitted to the court.

This is a case of formal cooperation in murder and hence intrinsically evil, for the lawyer consents in the will of the principal agent and intends the same purpose, euthanasia.[15] If, however, it was a question of a secretary taking down his client's request on paper, and not representing his client before the court or pleading his wish for euthanasia, the cooperation would be merely material, and for a good reason it would be permissible.[16]

(E) A member of the state legislature is approached by a representative number of constituents requesting that he introduce a euthanasia proposal before the assembly and argue in its favor. The representative does so, feeling that he has the right and duty to obey their wishes.

The state representative has done wrong. It may be that he does not will euthanasia *secundum quid,* yet *simpliciter* he does will it and is a formal cooperator in its enactment.[17]

(F) Judge Brown, in line with his office as judge, grants permission for a mercy killing when the legal requirements for the permit have been fulfilled. He asserts that in view of the enacted law he is compelled to grant euthanasia when the commission favors it. Hence he claims no responsibility for the mercy death. He believes also that morally speaking the case is similar to the judge granting a divorce.

Judge Brown may not grant euthanasia even at the cost of disobeying civil law or of resigning from his position as judge. By granting euthanasia he has become a formal cooperator in murder.[18]

---

[15] *Ibid.*

[16] H. Davis, *Moral and Pastoral Theology,* II, p. 342; B. Merkelbach, *Summa theologiae moralis,* II, nr. 492.

[17] H. Jone-U. Adelman, *Moral Theology,* p. 90.

[18] H. Davis, *Moral and Pastoral Theology* (New York: Sheed & Ward, 1943), I, p. 349. "No judge, under any circumstances, may pronounce judgment for what is essentially wrong."

It makes no difference whether that person is healthy or sick, or even requests euthanasia, the fact remains that the person is innocent and the judge may not send an innocent person to death.[19] St. Thomas has a few interesting remarks that affirm this:

> If the judge knows that a man who has been convicted by false witnesses is innocent, he must like Daniel examine the witnesses with great care, so as to find a motive for acquitting the innocent: but if he cannot do this he should remit him for judgment by a higher tribunal. If even this is impossible, he does not sin if he pronounce sentence in accordance with the evidence, for it is not he that puts the innocent man to death, but they who stated him to be guilty.[20]

It is evident that St. Thomas teaches that if necessary a judge may send an innocent man to death when objectively he appears guilty by reason of evidence. But there is no permission given a judge to send a man to death who is innocent even before the court. If Judge Brown grants an innocent person a civil right for euthanasia, he thereby sends an innocent man to death.

Judge Brown is also confused concerning the similarity between divorce and euthanasia. A divorce is not intrinsically evil, it is remarriage after divorce that is *in se* evil. In a divorce case the judge merely makes an official declaration that the state regards the civil effects of marriage as no longer existing. Such a declaration is morally indifferent.[21]

---

[19] F. Connell, *Morals in Politics and Professions*, p. 33. "The case is very different if a judge is called on to give a decision in favor of an action that is intrinsically wrong. Thus, in the years to come—particularly if we shall have on our hands a large number of persons physically and mentally incapacitated as a result of the war—the advocates of euthanasia may succeed in legalizing 'mercy killing.' Of course, law or no law, a judge would never be allowed to approve or decree such an act of murder."

[20] St. Thomas, *Summa Theologica*, II, II, Q. 64, Art. 6. "quod judex, si scit aliquem innocentem esse, qui falsis testibus convincitur, debet diligentius examinare testes, ut inveniat occasionem liberandi innoxium; sicut Daniel fecit: si autem hoc non potest, debet eum superiori relinquere judicandum: si autem nec hoc potest, non peccat secundum allegata sententiam ferens, quia ipse non occidit innocentem, sed illi qui eum asserunt nocentem."

[21] F. Connell, *Morals in Politics and Professions*, p. 31.

(G) A druggist sells a solution of poison which he knows will be used to administer voluntary euthanasia to an aged woman. He nevertheless sells it, thinking that if he does not some other druggist will.

The druggist, under the condition mentioned, may not sell the solution. He is a material cooperator in euthanasia and the facts presented do not justify his material cooperation.[22]

(H) A cancer patient living alone requests her physician to end her life in a quiet way. The physician knowing that such an act would be murder refuses, but he calls in a physician friend and asks him to administer euthanasia.

The physician did wrong in requesting his friend to put the patient to death. By this request he became a formal cooperator in murder.[23] In his position he could have offered sound counsel to the patient and perhaps changed her mind.

(I) A nurse is convinced that the quantity of the drug which the doctor has ordered for the patient will likely bring about immediate death. The nurse nevertheless carries out the doctor's orders and the patient dies.

Since the nurse had moral certainty about the lethal character of the dose, she sinned gravely. Her first duty was to call the physician's attention to this fact, and if it were clear to her that his previous orders were still to be carried out, she would be obliged to refuse. As a nurse she is not permitted to carry out an immoral order by reason of a doctor's command.[24] If the nurse were doubtful in the matter she should follow orders if he repeated the orders after she had told him of her suspicions. But for future guidance she should afterwards seek further information about the effects of the drug.[25]

(J) An orderly is told that the patient in his keeping has suicidal tendencies. The orderly is nevertheless careless and leaves poisoned

---

[22] J. Aertnys-C. Damen, *Theologia moralis*, I, p. 330, nr. 399.

[23] B. Merkelbach, *Summa theologiae moralis*, I, p. 400, nr. 489.

[24] M. Cronin, *The Science of Ethics* (New York: Benziger, 1939), I, p. 100.

[25] C. McFadden, *Medical Ethics for Nurses* (Philadelphia: Davis Co., 1946), p. 264. "If a nurse is in doubt about the morality of an operation, she may render any form of material assistance. But she should have the matter cleared up as soon as possible for her future guidance."

medicine within the patient's reach. Due to this negligence the patient succeeds in taking his own life.

The orderly has gravely failed in his duty and is therefore a *negative cooperator* in the suicide.[26] He has sinned against both charity and justice.[27]

(K) A policeman learns of the practice of mercy killing in a private home for incurable patients. He fails to act, however, for he learns that the local political "boss" is interested in the project, and is morally certain that in view of this fact no action on his part would be of any avail.

As the case stands the policeman does no wrong in failing to act. If he could end this affair, even though he might lose his position, he would be bound to act, but under the circumstances presented he commits no sin in remaining passive since he is sure that nothing would be done.[28]

(L) The same case is presented to the State Attorney General. He finds himself in a strong position to end this practice of murder, but he takes no action either for he seeks to be Governor after the next election and he will need the support of the local "boss" in question.

The State Attorney General is guilty of negative cooperation in murder.[29]

(M) A Catholic nun in a nationalized hospital serving as nurse, is ordered to assist in the administration of euthanasia to an old man. The assistance is remote and she believes that it is better to give this material assistance and remain on at the institution where she can do great good for the dying, than to refuse this material co-operation and be forced to leave. She is the only nun remaining at this time.

It is possible that in a case like the above there would be sufficient reason for remote material cooperation in euthanasia.[30] In extremely rare cases it might become licit for Catholic nuns or the laity to

---

[26] H. Noldin-A. Schmitt, *Summa theologiae moralis,* II, p. 117, nr. 116.
[27] J. Aertnys-C. Damen, *Theologia moralis,* II, p. 593, nr. 784.
[28] *Ibid.*
[29] *Ibid.*
[30] H. Davis, *Moral and Pastoral Theology,* I, p. 348.

place such cooperation in euthanasia as would be remote, namely, if otherwise these public institutions would eject them with subsequent great loss to religion and the welfare of souls.[31]

(N) A soldier is mortally wounded in battle and suffering great pain. The commanding officer orders a medical technician to end the man's life quickly as they must move forward and cannot care for the wounded man. The technician does as ordered.

The medical technician has acted immorally. The act of killing the wounded soldier was gravely illicit and murder.[32] Under no circumstance could he carry out the order even if the alternative would be his own immediate execution by firing squad and apparent public dishonor.

(O) A prisoner of war about to undergo decapitation is given the opportunity of choosing an easy death at his own hands. The official military court will appoint him, if he wishes, his own executioner. He takes the easy way out saying, "A lawful authority has condemned me to death and that same authority has appointed me legally my own executioner."

There is a probable opinion that a man may be his own executioner if appointed by a proper authority and empowered by a court as such. To take such authority upon himself simply because he is condemned to die or to presume the authority as given because the authority in charge makes such self execution easy is never lawful. In view of this the above case would seem to be permissible.[33]

(P) An old lady fearful of approaching death and its agony requests a sedative so that she may sleep away. The children having already attended to the spiritual preparation of their mother induce coma.

It is wrong under the circumstances to induce a coma even if it did not speed the old woman's death. The last moments of the dying are too important spiritually. This old lady was suffering a mental anguish at the thought of death and that was not sufficient reason

---

[31] *Ibid.*

[32] J. Aertnys-C. Damen, *Theologia moralis,* 1, p. 466, nr. 574. "Non licet occidere lethaliter vulneratos."

[33] B. Merkelbach, *Summa theologiae moralis,* II, p. 313, nr. 311; A. Koch-A. Preuss, *Handbook of Moral Theology,* V, p. 41.

for inducing coma.[34] Only very grave reasons would justify a heavy sedative so that the patient may sleep away.[35]

(Q) A lady is dying of tuberculosis and is suffering extreme pain. Sedatives seem no longer to quiet her agony but nevertheless she wishes to remain conscious till the end. After the Sacraments are given her and she is ready spiritually, her relatives ask the physician to give her a heavy sedative and let her sleep away. He does so and the lady dies while asleep.

A great wrong was done the patient. Though the patient was entitled to a heavy sedative according to the facts of the case, yet the will of the patient must always be respected in this matter. Sedatives may be given only with the permission of the patient.[36] Perhaps the patient wished to offer up the pains as penance.

(R) A cancer patient is in extreme pain and his system has gradually established what physicians call "toleration" of any drug, so that even increased doses give only brief respites from the ever-recurring pain. The attending physician knows that the disease is incurable and that the person is slowly dying, but because of a good heart, it is possible that this agony will continue for several weeks. The physician then remembers that there is one thing he can do to end the suffering. He can cut off intravenous feeding and the patient will surely die. He does this and before the next day the patient is dead.

The case involves the principle that an ordinary means of prolonging life and an extraordinary means are relative to the patient's physical condition.[37] Intravenous feeding is an artificial means of prolonging life and therefore one may be more liberal in application of principle. Since this cancer patient is beyond all hope of recovery and suffering extreme pain, intravenous feeding should be considered an extraordinary means of prolonging life. The physician was justified in stopping the intravenous feeding. He should make sure first, however, that the patient is spiritually prepared.[38]

(S) An old man, useless to others and a misery to himself, cuts

---

[34] H. Noldin-A. Schmitt, *Summa theologiae moralis*, I, p. 343, nr. 349.
[35] *Ibid.*
[36] S. La Rochelle-C. Fink, *Handbook of Medical Ethics*, p. 179.
[37] Cf. *supra*, p. ?.
[38] H. Noldin-A. Schmitt, *Summa theologiae moralis*, I, p. 343, nr. 349.

down on his food to the point of endangering his life. He believes that a man in his condition, at his age, may consider even daily meals an extraordinary means of prolonging life.

The old man is doing wrong. Eating is a natural means of prolonging life and according to the principle, a *per se* ordinary means of prolonging life.[39] It is possible that *per accidens* even eating could become for someone an extraordinary means, but given the circumstances in the present case there does not appear to be that *per accidens* condition that would allow the old man to consider eating an extraordinary means of prolonging his life.[40]

THE CATHOLIC PHILOSOPHY OF SUFFERING

It is difficult apart from faith to see how pain fits into a planned universe. To the worldly minded a cross is simply an evil they cannot escape. Pain and death they view as a sad triumph of a lower chemical order over the biological order which a "cruel and unreasoning nature," has inflicted upon them.[41] Even many good Christians do not understand the meaning of pain. They find a cross of physical suffering a contradiction to a good, holy and just God. It is Christ on that cross that solves the contradiction, but it takes the eyes of faith to see him there. And it is just this failure to see in their own lives the suffering Christ that makes pain so unbearable. The poor souls are undergoing pain without a purpose, without an end, and that indeed is unbearable. No wonder then so many request euthanasia.

Compare the difference between the happy deaths of cancer patients under the care of sisters in charitable homes, with the sad, hopeless, pitiful agony, if not contempt for God, so often found in similar cancer deaths in the public institutions. Two examples illustrate this point. Dr. Potter speaks of the fact that so many cancer patients have died "shrieking, groaning, and cursing till their breath failed."[42] And there is an account of the cancer patients in the Mother Alphonsa Home. The sister speaks of them as standing

---

[39] Cf. *supra*, p. 65.

[40] Cf. *supra*, p. 65.

[41] Cf. *supra*, Ch. 3, p. 22.

[42] Cf. *supra*, Ch. 3, p. 16.

their pain and offering it up to God as a penance for past sins and she adds: "They at times may be quite happy, and indeed, they are often willing to joke and laugh."[43] We must wonder as to why this great difference. It cannot be that those under the care of the sisters suffer less pain that those under the care of the finest state physicians and nurses. The most likely answer to the question is that the patients under the care of the sisters have learned the Catholic philosophy of suffering, and by seeing their suffering related to eternity and God's eternal plan, though they suffer much, they love even more, and by the grace of God their suffering has been raised to the dignity of a holy sacrifice.

This Catholic philosophy of suffering teaches the patient that suffering is not necessarily an evil. A thing is evil if it hinders one in the attainment of the purpose for which he exists. Suffering does not necessarily do this. Only sin is always and necessarily an evil.

> Ordinary observation of life shows that suffering may work in two ways. First it may be good for the sufferer. We know that a man who has never known suffering is soft and undeveloped. His character lacks substance. Immaturity clings about him. And not only do we find that this minimum of suffering is apparently necessary for man's proper development. We also find that really great suffering, if it has been dominated, has the power of enriching the character of the man or woman who has suffered. Suffering, if it ruins some characters, enriches others. It is not necessarily an evil, but may be an immense factor for good. What it is to be depends, for every man, on the way he accepts it. It lies in him to dominate it or be dominated by it.[44]

This thought of domination of suffering was set forth by one of the Church Fathers:

> It is not a blessed thing to be in the midst of suffering; but it is blessed to be victorious over it, and not to be cowed by the power of temporal pain.[45]

---

[43] J. Walsh, "Life Is Sacred," *The Forum*, XCIV (1935), p. 333.

[44] F. Sheed, *A Map of Life* (New York: Sheed & Ward, 1944), p. 102.

[45] St. Ambrose, *De officiis ministrorum*, II, 5 (P.L. 16,178) "non enim in passione, sed victorem esse passionis beatum est, nec frangi temporalis motu doloris."

Lite should be viewed as a period of testing and the suffering that arises in it is a part of that test. It is God that will measure the amount of suffering necessary for a man's perfection. The whole of life represents God's means of bringing a soul to its highest point of development. Some suffering is necessary; God knows how much each man needs; and it is by the suffering that cannot be legitimately avoided that God shows the measure of what is necessary.

The great test that the Christian must meet by pain and suffering demands that he voluntarily accept the suffering. To carry a cross in itself does not sanctify. All men carry a cross. It is the carrying of the cross in the proper spirit that makes one holy. The Catholic philosophy teaches one to do this. Men naturally flee from pain.

> Some souls, even pious souls, feel sometimes a certain sense of repulsion in hearing the word penance; it is the same with that of mortification which expresses the same idea. Whence does this feeling of repulsion come? It ought not to astonish us, for it has a psychological foundation. Our will necessarily seeks good in general, it seeks happiness, or that which seems to us to be such.[46]

When, however, the Christian realizes that this pain or suffering is not the end, but only the means to greater happiness even in this life, with God's grace he can join his will with that of God's. This act of the will to accept what all men flee from is in itself a triumph.

A Christian must know that the sign of his faith is the sign of the cross, and if he is to follow Christ he must take up the cross.[47] Then too, if Christians are other Christs, they must suffer, for the servant is not greater than the master.[48] Lastly because of many imperfections, faults and sins it is only fitting that Christians make reparation sometime, somewhere. It is rather the mark of a good and holy God that he permits so many of his children to undergo that suffering here on earth. Suffering is almost the greatest gift of God's love. For if we stop to think, we can never be like Him

---

[46] D. Marmion, *Christ the Life of the Soul* (London: Sands & Co., 1939), p. 186.

[47] St. Matthew, 16:24.

[48] St. John, 15:20.

in power or dignity, we can, however, become like Him in our suffering. In other words, by suffering we become God-like.

This is the Catholic philosophy of suffering. Anyone who lives by it could never request or support a request for euthanasia under any circumstances. Euthanasia has been proven morally evil. Yet, it is necessary to offer a "wounded world" something in its place, for one cannot displace euthanasia, one must replace it. It can be replaced by the Catholic philosophy of suffering.

BIBLIOGRAPHY

SOURCES

Acta Apostolicae Sedis, Commentarium Officiale, Rome, 1909 (Tomus I) sqq.

Catechism of the Council of Trent, translated into English by Rev. John A. McHugh, O.P. and Rev. Charles J. Callan, O.P., New York, Wagner, 1943.

Codex Iuris Canonici, Rome, 1917.

Corpus Juris, being a complete and systematic statement of the whole study of the law, ed. by W. Mack and W. Hale, New York: The American Law Book Co., 1920.

Corpus Scriptorum Ecclesiasticorum Latinorum, Academiae litterarum Caesareae Vindobonensis, Vindobonae, 1866 (Tomus I) sqq. (68 Vols., 1866-1936) (CSEL).

Denziger, Henr., et Bannwart, Clem., Umberg, Joan., *Enchiridion Symbolorum, Definitionum, et Declarationum de Rebus Fidei et Morum,* Editio 21-23, Friburgi Brisgoviae: Herder, 1937 (DBU).

Encyclopedia Britannica, 11 ed., 1910-1911.

Mansi, Joannes, *Sacrorum Conciliorum Nova et Antiquissima Collectio,* 53 vols., in 59, Florentiae, Paris, Arnheim, et Leipzig, 1901-1927.

Migne, J. P., *Patrologia Cursus Completus, Series Graeca,* 161 vols., Paris, 1858-1866.

————, Patrologia Cursus Completus, Series Latina, 221 vols., Paris, 1844-1864.

REFERENCE WORKS

Aertnys, Josephus, et Damen, Cornelius, *Theologia moralis secundum doctrinam,* S. Alphonsi de Ligorio, 18-21 ed., 2 vols., Turin: Marietti, 1944.

Alphonsus, Liguori, St., *Theologia Moralis,* ed. L. Gaude, 4 vols., Rome: 1905-1912.

————, Homo Apostolicus, ed. nova, ed. Saraceno, Turin: Marietti, 1890.

Aquinas, St. Thomas, *Summa theologiae,* Ed. Institute of Medieval Studies, Ottawa, 1942.

Antonelli, Joseph, *Medicina Pastoralis,* 5th ed., Rome: Pustet, 1932.

Augustine, St., *De Civitate Dei,* ed. (CSEL).

Arregui, A., *Summarium Theologiae Moralis,* Ed. XII, Bilboa, 1934.

Aristotle, *The Basic Works of Aristotle,* ed., Richard McKeon, New York: Random House, 1941.

Ballerini, A.-Palmieri, D., *Opus theologicum morale*, 7 vols., Ed. III, Paris, 1889-1893.

Bellacosa, L., *Manuale di theologia morale*, Naples, 1837.

Bellarmine, Robert, St., *Roberti Bellarmini Opera Omnia*, Naples: Guiliano, 1858.

Billuart, F. C., *Summa Sancti Thomae hodiernis Academiarum moribus accomodata*, Ed. Nova, Paris: 1848.

Bonacina, M., *Operum de morali theologia*, Venice, 1687.

Bonal, A., *Institutiones theologiae*, Ed. VIII, Toulouse, 1893.

Bouquillon, T., *Institutiones theologiae moralis fundamentalis*, Bruges: Beyaert-Defoort, 1873.

Bonnar, A., O.F.M., The Catholic Doctor, New York: P. J. Kenedy, 1930.

Busenbaum, H., S.J., *Medulla theologiae moralis*, Turin, 1848.

Bancroft, *The Native Races*, New York: Harper, 1883.

Capellman, Dr. C., *Medicina Pastoralis*, Ed. VII, Aquisgrani, 1890.

Catholic Encyclopedia, The, XV Vols., Index and II Sups., New York: 1907-1922.

Conway, B. L., *The Church and Eugenics*, New York: Paulist Press, 1946.

Coppens, C., and Spalding, H., *Moral Principles and Medical Practice*, new and enlarged edition, New York-Cincinnati: Benziger, 1921.

Crolly, G., *Disputationes theologicae de justitia et jure*, Dublin: Gill and Sons, 1877.

Cronin, M., *The Science of Ethics*, 2nd ed., New York: Benziger. 1939.

D'Annibale, J., *Summa theologiae moralis*, Ed. V, Rome, 1908.

Davis, H., S.J., *Moral and Pastoral Theology*, 3 ed., 4 vols., New York: Sheed and Ward, 1938.

De Lugo, John, S.J., *De justitia et jure*, Ed. Novissima, Tom. I, Lyons, 1670.

Dictionnaire de théologie Catholique (publié sous la direction de A. Vacant et de E. Mangenot), Paris, 1903.

Donovan, D. A. Q. Cist., *Compendium theologiae moralis*, St. Louis, 1897.

Dorsay, J., "Omaha Sociology," *Annual Reports of the Bureau of American Ethnology*, 1882.

Elbel, B., O.M., *Theologia moralis*, Augustae Vindelicorum et Oeniponti, 1751.

Farrell, W., O.P., *A Companion to the Summa*, 4 vols., New York: Sheed and Ward, 1939.

Farges, A. et Barbedette, D., *Philosophia scholastica*, 2 vols., Paris: 1937.

Ferreres, J., S.J., *Compendium theologiae moralis*, Ed. XIII post Codicem, Barcelona: 1925.

Five Great Encyclicals, The Paulist Press, New York: 1939.

Fallen, V., S.J., *Eugenics*, Translated by E. C. Messenger, New York: Benziger, 1923.

Genicot, E.-Salsmans, I., *Institutiones theologiae moralis*, 13th ed., 2 vols. Brussels: L'Edition Universelle, S.A., 1936.

Gould, G., *Medical Dictionary*, Philadelphia, 1929, Ed. II.

Gury, P., S.J., *Compendium theologiae moralis*, Ratisbon: 1868.

Gury, P.-Ballerini, A., *Compendium theologiae moralis*, 2 ed., Rome: 1869.

Henno, F., O.M., *Tractus moralis in Decalogi Praecepta*, Ed. III, Turin: 1711.

Hinman, F., "Power Over Life and Death," *Journal Nervous and Mental Diseases*, IC (1944).

Howard, R., "Taking Life Legally," *Magazine Digest*, XC (1947).

Iorio, T., S.J., *Theologia moralis*, Ed. VI, Naples, 1939.

Johnson, D., *Facing the Facts about Cancer*, New York: Committee Press, 1947.

Juliano, J., S.J., *Manuductio ad theologiam moralem*, Padua, 1787.

Kenrick, F. P., *Theologia moralis*, Mechlinae: 1860.

Koch, A.-Preuss, A., *Handbook of Moral Theology*, Ed. III, St. Louis, 1926.

Kennedy, F., "Unfit to Live," *American Journal of Psychiatry*, IC (1942).

Kidd, D., *The Essential Kafir*, London: Black, 1925.

La Croix, J., *Theologia moralis*, Ravenna, 1755.

La Rochelle, S., O.M.I.,, and Fink, C., M.D., C.M., *Handbook of Medical Ethics* (Tr. from 4th French Edition by M. E. Poupore), Westminster: Newman, 1943.

Laymann, P., S.J., *Theologia moralis in Libri V. Partita*, Venice, 1638.

Lehmkuhl, A., S.J., *Theologia moralis*, Ed. XII, Friburg, 1914.

Lessius, L., S.J., *De justitia et jure*, Ed. IV, Antwerp: 1617.

Loiano, C., O.M.Cap., *Institutiones theologiae moralis*, Turin, 1935.

Laski, H. J., *Studies in the Problem of Sovereignty*, New Haven: Yale, 1917.

Liddell, H.-Scott, R., *A Greek-English Lexicon*, 8 ed., Oxford: Clarendon Press, 1897.

Linderman, F., *Red Mother*, New York: Longmans, Green, 1932.

Laertius, *Lives of Eminent Philosophers* X, LCL, Tr. by Hicks, New York: Putnam's, 1925.

Literary Digest (March 1925).

Merkelbach, H., O.P., *Summa theologiae moralis*, Ed. III, Paris: Desclée, 1938.

Michels, A., "Mutilation," *Dictionnaire de théologie Catholique*, Vol. X, Col. 2569 (publié sous la direction de A. Vacant et de E. Mangenot), Paris 1903.

Molina, L., S.J., *De justitia et jure*, Venice: 1611.

More, T., St., *Utopia*, London: Bell and Stone, 1910.

Morino, J., C.M., *Theologia moralis*, Ed. IX, Naples, 1922.

Muller, E., *Theologia moralis*, Ed. Ignatius Seipel and Joseph Ujcic, Ratisbon: 1938.

Noldin, H.-Schmitt, A., S.J., *Summa theologiae moralis*, ed. XXV, Ratisbon: 1940.

New York Times Index (1938-1947).

New English Dictionary on Historical Principles, Oxford, Clarendon Press, 1897.

O'Malley, A., *The Ethics of Medical Homicide and Mutilation,* New York, 1919.

Pius XI, Pope, Encyclical Letter "Casti Connubii," *Acta Apostolicae Sedia,* Vol. 22, Rome, 1930.

Pius XII, Pope, Encyclical Letter "Mystici Corporis," *Acta Apostolicae Sedis,* Vol. XXXV, Rome: 1943.

Prümmer, D., O.P., *Manuale theologiae moralis,* ed. III, Friburg: 1923.

Proposed Bill to Legalize Euthanasia, Prepared by the Euthanasia Society, of N. Y.

Plutarch, *Plutarch's Lives,* XVI (LCL, I, Tr. by Perin), London: Macmillan, 1914.

Reiffenstuel, A., O.M., *Theologia moralis,* Munich, 1702.

Rohling, A., *Medulla theologiae moralis,* St. Louis, 1875.

Ryan, J. A., *Human Sterilization,* Washington: National Catholic Welfare Conference, 1936.

———, *Moral Aspects of Sterilization* (Problems of Mental Deficiency nr. 3.), first and revised editions, Washington: National Catholic Welfare Conference, 1930.

Renard, H., S.J., *The Philosophy of Being,* Milwaukee: Bruce, 1947.

Ratzel, F., *The History of Mankind,* London: Macmillan, 1897.

Sabetti, A.-Barrett, T., *Compendium theologiae moralis,* ed. XXXIV, New York, 1939.

Slater, T., Moral Theology, New York: 1908.

Sporer, P., O.M., *Theologiae moralis Decalogum,* Salzburg: 1686.

Suarez, F., *Opera Omnia,* ed. by Charles Berton, Paris: 1858.

Sylvius, F., *Commentarii in totum secundam secundae S. Thomae Aquinatis,* Venice, 1726.

Simmons, L., *The Role of the Aged in Primitive Society,* London: Oxford, 1945.

Stevenson, R., *In the South Seas,* New York: Scribner's, 1918.

Thompson, J., C.M., *Lectures on Medical Ethics,* Sydney, 1933.

Ubach, J., S.J., *Compendium theologiae moralis,* Friburg, 1926.

Vermeersch, A., S.J., *Theologica moralis principis, responsa, consilia,* Rome: 1933.

Vaughan, J. M., "Blood Transfusion," British Medical Journal, London, 1931.

Vives, Card. J., *Compendium theologiae moralis,* ed. VII, Rome: 1902.

Wouters, L., C.Ss.R., *Manuale theologiae moralis,* Bruges, 1932.

SELECTED ARTICLES AND MINOR REFERENCES

Anderson, S., "Dinner In Thessaly," *Forum* (1936).

Brice, J., "Reflections on Euthanasia," *Journal of Nervous and Mental Disease* (July, 1936).

Cabot, H., M.D., "Euthanasia," *New York Herald-Tribune* (Jan. 17, 1940).

Carroll, G., "Kristin," *Red Book* (April 1939).

Cronin, A. J., "Inheritance," *Cosmopolitan* (April 1939).

Doane, J., M.D., "Our Service to the Dying," *Journal of Nursing* (Nov. 1938).

Euthanasia Society of New York, *Merciful Release*, Euthanasia Press, 1947.

Halton, M., M.D., "No Doctor Should be a Mercy Killer," Physical Culture (April 1936).

Kosley, V., R.N., "As Life Ebbs," *Journal of Nursing* (Nov. 1938).

Kennedy, F., "Euthanasia," *Collier's* (May 1939).

Lennox, W., "Should They Live," pamphlet issued by William G. Lennix, M.D., Autumn, 1938.

Little, C., "Let Us Face Death," *Scribner's* (June 1931).

Philbrick, I., M.D.. "Euthanasia," Radio Address.

Roberts, H., M.D., *Euthanasia and Other Aspects of Life and Death*, Constable, London: 1936.

Tollemache, L., *Stones of Stumbling*, Hodgson and Son, London: 1887.

Walsh, J., "Life Is Sacred," *Forum* (Dec. 1935).

Wolbarst, A., "The Right to Die," *Forum* (Dec. 1935).

———, "The Doctor Looks at Euthanasia," *Medical Record* (April, 1939).

INDEX

Abandonment, 4, 6, 11, 12.
Abortion, 51.
Abraham, 29.
Acta, 45, 52.
Administration of euthanasia, 28.
Aertnys-Damen, 33, 36, 37, 38, 39, 54, 69, 70.
Aggressor, killing of, 37, 38.
Agony, 63.
Alphonsus, 36, 37.
Ambrose, 49, 74.
American Advisory Council, 21.
Amputations, 65.
Anaesthetics, against will, 72.
  to escape agony, 71.
Anderson, 15.
Animals, killing of, 31.
Anselm, 49.
Anthropology, 4, 5, 6, 11.
Application for euthanasia, 25, 26.
Aquinas, 32, 34, 36, 37, 38, 49, 58, 68.
Arawak tribe, 6.
Aristotle, 7, 8, 31, 33.
Athanasius, 49.
Augustine, 29, 31, 32, 34, 48, 57.
Austria, euthanasia in, 57.
Auxere, Council of, 50.

Ballerini-Palmieri, 42.
Bancroft, 5.
Barbarism, 54.
Barrett, 7.
Basil, 49.
Bede, 49.
Benedict, 49.
Bertke, 60, 62.
Bibliography, 77, 78, 79, 80, 81.
Billuart, 42, 53.
Biographical note, 82.
Bodily integrity, 45.
Bonnar, 2.
Braga, Council of, 50.
Brown, 56.
Burial of suicides, 50, 51.
Bushman tribes, 4.

Cancer, 16, 17, 22.
Canon Law, on suicide, 51.
  on duel, 51.
  on abortion, 51.

Capital punishment, 34.
Catechism of Council of Trent, 50.
Catholic philosophy of suffering, 73.
Charity, 45, 66.
Chrysostom, 49.
Codex Iuris Canonici, 51.
Coma, 71, 72.
Combatants, killing of, 36.
Commandment, fifth, 60.
Common good, 43.
Comstock, 21, 22.
Connell, 68.
Conservation of life, 64, 65.
Cooper, 60.
Cooperation in euthanasia, 66, 67, 69, 70.
Court Committee, 27.
  jurisdiction, 25.
Cronin, 33, 36, 37, 38, 58, 69.
Crow Indian, 4.
Culture, primitive, 4, 5, 6.
  Greek, 7, 8.
  Roman, 9, 10.
  Northern Europe, 11, 12.
Cunningham, 45.
Custom of West, 47, 48, 49, 50, 51, 52, 53, 54, 55, 56, 57.
Cyprian, 49.

Daniel, 53.
Davis, 33, 36, 39, 67, 70.
Death's agony, 63.
Decalogue, fifth Commandment, 60.
Definitions, 25.
DeLugo, 36, 37, 40, 42, 43, 53, 66.
Deuteronomy, 29, 30.
Direct voluntary act, 38, 39.
Disease incurable, 17, 43.
Divorce unlike euthanasia, 68.
Doctors of Church, 53.
Dominium plenum et utile, 43.
Dominion, supreme, 40, 41, 42, 46, 66.
Dorsey, 4.
Druggist, 69.
Duelling, 51.
Dunn, 14.

Encyclicals, 52, 45.
English law on suicides, 50.

82

BIOGRAPHICAL NOTE

Joseph V. Sullivan was born in Kansas City, Missouri, August 15, 1919. He received his elementary education at St. Vincent's parochial school in Kansas City, and his high school education at De La Salle Academny in Kansas City. He entered the Kansas City Diocesan Seminary in 1937 for college work and in September of 1940 he began his philosophical studies at the St. Louis Preparatory Seminary in St. Louis, Missouri. In 1942 he began the study of theology at the Catholic University of America, and in 1946 he was ordained to the priesthood. June 12, 1946, he received the degree of Licentiate in Sacred Theology. In September of 1946 he returned to the Catholic University of America for graduate study.

www.ingramcontent.com/pod-product-compliance
Lightning Source LLC
Chambersburg PA
CBHW060152290526
45789CB00003B/1006